Bo

There are nine other titles in this series:

The Law and Practice Relating to Pollution Control In

Denmark
France
Federal Republic of Germany
Greece
Ireland
Italy
The Netherlands
The United Kingdom

The Law and Practice Relating to Pollution Control in the Member States of the European Communities: A Comparative Survey

The series will be updated at regular intervals. For further information, complete the enclosed postcard and send it to:
Graham & Trotman Limited
Sterling House
66 Wilton Road
London SW1V 1DE

All the titles in the series were prepared by

Environmental Resources Limited
79 Baker St, London W1M 1AJ (Tel. 01-486 8277; Tx. 296359 ERL G)

or

he Commission of the European Communities,
irectorate-General Environment, Consumer Protection
d Nuclear Safety, Brussels

The Law and Practice Relating to Pollution Control in Belgium and Luxembourg

The Law and Practice Relating to Pollution Control in Belgium and Luxembourg

Second Edition

Prepared by

Professor L. P. Suetens
University of Louvain (K. U. Leuven)
Member of the Council of State

Dirk Soetemans
University of Louvain (K. U. Leuven)

for

Environmental Resources Limited

Published by
Graham & Trotman
for
The Commission of the European Communities

Published in 1982 by

Graham & Trotman Limited
Sterling House
66 Wilton Road
London SW1V 1DE

for

The Commission of the European Communities,
Directorate-General Information Market and Innovation,
Luxembourg

EUR 7732

British Library Cataloguing in Publication Data

Suetens, L.P.
   The law and practice relating to pollution
   control in Belgium and Luxembourg.—2nd ed.
   1. Pollution—Law and legislation—Belgium
   2. Pollution—Law and legislation—
   Luxembourg
   I. Title   II. Soetemans, Dirk
   III. Environmental Resources Ltd.
   IV. Commission of the European Communities
   344.9304'463

   ISBN 0-86010-310-2

The views expressed in this publication are those of the authors, and should not be taken as reflecting the opinion of the Commission of the European Communities.

LEGAL NOTICE

Neither the Commission of the European Communities nor any person acting on behalf of the Commission is responsible for the use which might be made of the following information.

Printed in Great Britain by
Robert Hartnoll Limited, Bodmin, Cornwall

Summary List of Contents

LUXEMBOURG

Preface

This volume is part of a series prepared in the performance of a contract between the Commission of the European Communities and Environmental Resources Limited (ERL). ERL is a consulting organisation specialising in environmental research, planning and management.

In 1976 a first series was published covering the, then, nine members of the Community. The purpose of those volumes was to explain the law and practice of pollution control in each of the Member States and to provide a summary comparing all the countries in a separate comparative volume.

Since that time many changes in legislation arising from both national and Community-wide initiatives have occurred. ERL was therefore asked to prepare a new series providing an up-to-date review of the law and practice relating to pollution control in the Member States of the European Community.

The series comprises nine volumes concerning the law and practice in the Member States:

Belgium and Luxembourg   Ireland
Denmark   Italy
France   The Netherlands
The Federal Republic of Germany The United Kingdom
Greece

and a summary comparative volume.

The aim of this new series, as in the first, is to provide a concise but fully referenced summary of the letter of the law, and a discussion of its implementation and enforcement in practice. Proposals for new legislation which has been drafted but not yet passed are outlined. Where laws have been introduced to comply with Community-wide requirements this is noted.

The publication has two principal objectives:

to enable the reader to study in outline the provisions in any one Member State; and

to enable a direct comparison between different Member States.

To facilitate comparison between the national reports, each is indexed following a standard format (the Classified Index) to enable easy reference to the relevant sections of each report.

Presenting a nation's laws accurately in summary form is always a difficult task. There is a danger that, out of context, they may be misunderstood. We have therefore tried to give, in the first section of each report, some of the constitutional, legal and administrative background.

A further danger lies in translation. Although in the English texts we have tried to prepare as accurate a translation as possible, only the authors' original texts in their native languages carry their full authority. These texts are also being published in the individual Member States.

The statement of law in each volume is correct to at least 30 June 1981; in some cases more recent revisions have been included during the period of preparation for publication.

The series will be updated at regular intervals; to receive further details readers should complete the enclosed postcard and send it to the publisher.

ERL would like to acknowledge and express its thanks for the contributions from the national authors and for their cooperation in the preparation of the series.

Finally, ERL also acknowledges the assistance provided by many agencies, which have freely given information and advice, and the help and guidance given by Monsieur Claude Pleinevaux, Mr Grant Lawrence and other members of the Directorate of Environment, Consumer Protection and Nuclear Safety of the Commission of the European Communities.

1982   Environmental Resources Limited
                                                                   London

Detailed List of Contents

BELGIUM

CONTENTS

LUXEMBOURG

CONTENTS

ACKNOWLEDGEMENTS

It is my welcome duty to express my gratitude to all those who have helped me in various ways to complete this book.

As for the part on Belgium, I want to thank in particular Mr Dirk Soetemans, who wrote the first draft of seven chapters, Mr Hubert Bocken, author of the chapter on compensation for damage to the environment, and Mrs Geneviève Pal, who wrote the first draft of the chapter on disposal of waste.

As for the part on Luxembourg law, I am under a great obligation to Mr Léon Rinnen and Mr Joseph Peschon (Department of the Environment), and to Mr Guy Glodt (Council of State), who have given me invaluable information.

Many thanks to Mrs Agnes Plancquaert-Dillaerts, who prepared the manuscript for publication with skill and good humour.

My gratitude, finally, to Ms Karen Raymond of Environmental Resources Limited, London, for her support and patience.

The blemishes that remain in this book are clearly my own. I apologise in advance for them.

Professor Dr L. P. Suetens

The Law and Practice Relating to Pollution Control in Belgium

1
Political and Administrative Institutions

1.1 GENERAL

1.1.1 Introduction

Belgium is a representative democracy in the form of a constitutional monarchy.

Articles 25–31 of the Belgian Constitution are devoted to the theory of the 'separation of power' as a condition of political freedom, dividing the public functions into three distinct and independent groups: legislative power, executive power and judicial power.

1.1.2 Communities and Regions

Since the Constitution was amended in 1970, Belgium has been divided into *three Communities:* the French Community, the Flemish Community and the German Community (article. 3, *ter* of the Constitution). The French and Flemish Communities have a legislative organisation at their disposal; the German Community—80,000 inhabitants—has a council which issues regulatory decrees. The Communities have an executive organisation of their own.

The amendment to the Constitution mentioned above also created *three Regions:* the Flemish Region, the Walloon Region and the Brussels Region. To date, only the Flemish and Walloon Regions have a legislative organisation at their disposal.

Each Region has its own executive, but the executive for the Brussels Region consists only of a ministerial committee within the national government.

In the north of the country, the Flemish Region coincides with the Dutch Community, but the Dutch Community is also present in the Brussels Region; in Brussels and the south of the country the Brussels Region includes part of the French Community—which is the majority there—and the Walloon Region includes the largest part of the French Community, but also the German Community.[1]

1.1.3 Legislative powers

Legislative power is therefore split between the national parliament—that is, the Chamber of Deputies and Senate—on the one hand, and the Flemish Council, the Council of the French Community and the Walloon Regional Council on the other hand.

The *National parliament* legislates on all matters which have not been expressly transferred to one of the three other legislative bodies.

In other words, as a general rule, legislative power is exercised by the Chamber and the Senate; the other councils only have the responsibilities expressly conferred upon them by the Constitution.

National legislative power is exercised jointly by parliament and the King who can present draft laws to the Chamber of Deputies or the Senate. In addition to this, he ratifies national laws.

The community and regional councils rule by means of decrees which have the force of law on matters within their competence. Decree power is exercised jointly by the Council and the Executive.

1.1.4 Councils

The *Flemish Council* is responsible for matters covered by article 59, *bis* of the Constitution—that is, cultural and 'personalisable' matters (see 1.1.5)—and within the Flemish Region also carries out the responsibilities attributed to the regional bodies with regard to 'regional' matters. There is therefore only one council for the Flemish Community and the Flemish Region.

The *Council for the French Community* is responsible for cultural and 'personalisable' matters; the *Walloon Regional Council* is responsible for 'regional' matters. It is provided for that the Council for the French Community and the Council for the Walloon Region can decide by common accord, by decree voted by a two-thirds majority, that the Council and the Executive of the French Community shall carry out in the Walloon Region the responsibilities of the regional bodies concerning matters under article 107, *quater* of the Constitution.

From the last integral renewal of the legislative chambers, the councils are made up of deputies and directly elected senators.

1.1.5 Responsibilities of councils and regions

The responsibilities transferred to the councils are listed in articles 4–6 of the special law on institutional reform, dated 8 August 1980.[2]

Article 4 lists cultural matters; these are mainly culture in its strictest sense: teaching,[3] cultural co-operation, the use of language, education, as well as reorientation and professional retraining.

Article 5 lists 'personalisable' matters; these are mainly matters concerning public health and hygiene, the assistance of individuals, families and services, as well as youth protection matters.

Article 6 lists regional matters; these are mainly town and country planning, the protection of the environment, rural improvements and nature conservation, housing, water policy, certain aspects of economic, energy and employment policy.

Article 7 also transfers responsibilities to the Regions regarding the organisation of procedures and the manner in which central control of the provinces, communes, conglomerations and federations of communes is exercised.

Articles 8–16 specify certain responsibilities of the councils:

creation and organisation of public corporations;

adoption of provisions concerning the infrastructure;

implied powers;

ability to establish punishments;

the transfer to the Communities and Regions of State property which is required in the carrying out of their functions;

budget vote;

power to contract loans and to initiate public borrowing;

assent of community councils on certain international treaties.

1.1.6 Executive powers

Executive power is exercised on a national level by the national government. There is also a 'Flemish Executive', an 'Executive for the French Community' and a 'Walloon Regional Executive'.

The members of the Executives are elected by the Councils, but mandates for Executive members are now divided proportionately among the political groups which make up the Council. The Executive appoints a President from its members.

The Executives are accountable to the Councils. The Council can pass a motion of no confidence in the Executive or in one or several of its members. This motion is only valid if it proposes a successor to the Executive or members in question.

1.1.7 Judicial powers

Judicial power is exercised by the courts and tribunals.[4]

There is one justice of the peace in each judicial canton. In certain cantons there is also a police tribunal with responsibilities in police matters; in other cantons the functions of the police tribunal judge are carried out by the justice of the peace.

In principle, the justice of the peace handles all claims which do not exceed 50,000 francs. The police tribunal handles petty offences.

There is a district court, a general court of first instance, a labour court and a commercial court in each district.

The district court rules on jurisdictional conflicts between the court of first instance, the labour court and the commercial court.

The court of first instance is made up of three sections: civil court, criminal court and youth court.

The civil court is the ordinary court in civil matters; the police court deals in the first resort with all offences for which the punishment exceeds seven days' imprisonment and a fine of 25 francs; the youth court deals with summonses relating to minors under 18 years of age; the commercial court handles litigation of a commercial nature; the labour court is responsible for resolving individual conflicts relating to work.

The civil court also handles appeals against judgements delivered in the first resort by the justice of the peace; the police court handles appeals against judgements delivered in the first resort by the police tribunal.

1.1.8 Appeal courts

As a result of the reform of the Constitution on 11 June 1970, there are five appeal courts in Belgium, having equal status. The courts of appeal handle decisions made in the first resort by the court of first instance and the commercial court.

There are also five higher labour courts handling appeals against decisions granted in the first resort by the lower labour courts.

1.1.9 Court of cassation

There is only one court of cassation for the whole of Belgium (article. 95 of the Constitution).

The Court of Cassation handles decisions granted in the last resort which are referred for a decision on whether or not they contravene the law or violate certain forms which must be observed.

The Court of Cassation does not deal with the entire case; if it annuls a decision given by a lower ranking judge, the case must be referred to another court which will re-examine it completely.

The Court of Cassation also decides on conflicts of jurisdiction when a judicial court and an administrative tribunal are in dispute over the handling of a case.

1.1.10 Public Prosecutors

Article 101 of the Constitution envisages the creation of a Public Prosecutor's Office. Public Prosecutors are the executive officials at the courts and tribunals. They are responsible for implementing public action and for pursuing offenders with regard to police matters; in civil matters they intervene when such intervention is required to maintain public order.

1.1.11 Administrative jurisdictions

The legislator can establish other jurisdictions in addition to the ordinary courts and can confer upon them the responsibility for judging disputes relating to political rights.

In fact, a great number of administrative jurisdictions with special responsibilities have been created. On this subject, one author has commented: 'This jurisdictional mosaic, made up of different bodies created from time to time, does not correspond to any overall plan'.[5]

1.1.12 Local bodies

Belgium is divided into nine provinces; there are 596 communes.

'The province is both a territorial area of the country and an autonomous political entity responsible for the management of provincial interests. As a part of the State, the province participates in the administration of the Kingdom. But provincial power is also, and above all, an autonomous power'.[6] The same applies to the communes.

Interests which are exclusively communal or provincial are controlled by the communal or provincial councils, according to the principles laid down in the Constitution. These principles are (article. 108):

(1) the direct election of members of provincial and communal councils;

(2) the transfer to the provincial and communal councils of responsibility for all matters of provincial and communal interest, without prejudice to approval of their acts;

(3) the decentralisation of transferred powers towards the provincial and communal institutions;

(4) public access to sessions of the provincial and communal councils;

(5) public access to budgets and accounts;

(6) intervention by the controlling or legislative authority to prevent the law being violated or public interest being harmed.

The Constitution also envisages the creation of conglomerations and federations of communes but in fact only one conglomeration—the Brussels conglomeration—has so far been created and that is only provisional.

1.1.13 Sources of law

The Constitution is the supreme law. The national legislator has statutory sovereignty in all matters which have not been removed from his sphere of activity by the Constitution; these are *laws* in the organic sense.

Decrees issued by the community councils and regional councils have the force of law. 'The King issues the necessary *regulations and decrees* for the implementation of laws, without having the power to suspend the laws themselves or to dispense with their implementation' (article. 67, Constitution). The Royal Decree is therefore only a derivative and secondary provision.

International treaties are made by the King; however, trade treaties and treaties which could impose duties upon the State or are directly binding on Belgian individuals can only be implemented after they have received the assent of the Chamber and the Senate. International treaties—as well as EEC law—shall be implemented either by the national government, or by the Executive of the Communities or Regions. In a decision of 27 May 1971, the Court of Cassation upheld the principle of the supremacy of international law which applies to internal law even following the ratification of the treaty.[7]

Rules issued by the European Communities form an integral part of the relevant national law.

The provincial councils issue *provincial regulations* of internal administration and police orders.

The communal councils issue *communal regulations* of internal administration and communal police orders; these are aimed at maintaining public safety, peace and salubrity in the commune (article 3, heading XI, of the Decree of 16–24 August 1790 on judicial organisation).

Communal and provincial regulations cannot apply to matters already controlled by laws or regulations of the general administration. They are automatically repealed if general administration laws or regulations are issued subsequently. In addition, provincial regulations and orders cannot apply to matters which are the responsibility of the commune.

Judges cannot give a ruling by means of general or regulatory provisions on cases submitted to them (article 6, Judicial Code). Even though *decrees and judgements* are only relative in scope (excluding repeal decrees issued by the Council of State which are binding on all), they have an undeniable power of persuasion.

1.2 POLLUTION CONTROL

1.2.1 Introduction

In principle, the pollution control is one of the responsibilities transferred to the Regions (see 1.1.5). This principle should nevertheless be specified and examined.

Article 6, s.1, II, of the Law of 8 August 1980 states that 'As far as the environment is concerned:

(1) the protection of the environment, in respect of general and sectoral legal provisions;

(2) the removal and treatment of waste;

(3) the control of dangerous, dirty and noxious establishments, with the exception of provisions relating to protection'

are all regional matters.

Let us examine this statement further, point by point.

1.2.2 The protection of the environment

By inserting the words 'protection of the environment', the legislator intended to transfer the responsibility for the fight against certain nuisances, particularly air and water pollution and noise, to the Regions.

But environmental protection thus defined is controlled by three national laws:

the Law of 28 December 1964 relating to the control of atmospheric pollution;

the Law of 26 March 1971 on the protection of surface waters against pollution;

the Law of 18 July 1973 relating to the control of noise.

There are, in addition, international provisions such as, for example, acoustic standards for aviation fixed by the International Convention on Civil Aviation.

The Regions therefore have no power on a legislative level. Quite the contrary; by 'legal provisions' is meant the law and its implementing decrees. The regions apply the national and international 'legal provisions'. In this context they can impose stricter requirements than do the general and sectoral provisions, but they cannot amend legal or regulatory provisions.

1.2.3 Waste

Henceforth the Regions can legislate independently, by decree, on matters concerning the removal, treatment and depositing of waste, including toxic waste.[8]

1.2.4 Classified establishments

The Regions are responsible for the *external control of establishments classified as dangerous, dirty or noxious,* that is, control 'which is strictly linked to land use planning'.[9]

In the present state of the legislation and regulations it is already possible, with regard to the creation and operation of a noxious establishment, to distinguish between:

(1) the provisions regarding town planning and land use planning, i.e. the obligation to apply for a building permit for an establishment of this kind;

(2) the provisions contained in the General Regulation for Protection at Work, i.e. the obligation to apply for an operating licence.

Two aspects are taken into consideration before an operating licence is granted:

(a) the nuisance which such an establishment may create for the environment (damage to the environment and inconvenience for the inhabitants—aspects which are therefore related to land use planning and the environment);

(b) the inconvenience which the operation may create for the workers in such an establishment (aspects relating to safety and hygiene at work and to medicine at work).

The responsibilities transferred to the regions are town planning and land use planning—responsibilities envisaged under (1) above—and external control envisaged under (2)(a); internal control envisaged under (2)(b) remains a national matter.

1.2.5 Wastewater

While it is undeniable that wastewater treatment only constitutes one aspect of 'environmental protection', the Law of 8 August 1980, an institutional reform, contains a specific provision on the subject.

In principle, the Region is responsible for the purification of wastewater and for drainage; this responsibility does not, however, include:

(a) the establishment of the general and sectoral conditions for the discharge of wastewaters;

(b) the establishment of the constituent elements for the calculation of tariffs for treatment of industrial effluent;

(c) the fixing of a subsidy rate for industrial businesses.

In fact, the autonomous legislative regulatory and power of the regions on this subject is only very marginal. There is, however, one small consolation: the regional executives must be 'associated' with the elaboration of national preliminary draft laws and regulations on subjects under (a) and (b).

1.2.6 Responsibilities

On 1 January 1981, the responsibilities on the subject of the environment were defined as follows:

(a) *Flemish Region* (Royal Decree of 18 January 1982, *Moniteur belge*, 5 February 1982):

Environmental and water policy— J. Lenssens, Minister of the Flemish Community;

Town and country planning, rural improvement and nature conservation—P. Akkermans, Minister of the Flemish Community.

(b) *Walloon Region* (Royal Decree of 12 March 1982, *Moniteur belge*, 12 May 1982):

Environmental policy with the exception of industrial treatment of waste, and water policy, rural improvement and nature conservation—V. Féaux, Minister of the Walloon Region;

Industrial treatment of waste—Ph. Busquin, Minister of the Walloon Region.

(c) *Brussels Region:*

Water policy and town and country planning—Mrs C. Goor-Eyben, Secretary of State for the Brussels Region;

Environmental policy, waste treatment, water pollution control and nature conservation—Mrs A. M. Neyts-Uyttebroeck, Secretary of State for the Brussels Region.

(d) Matters not transferred to the regions—F. Aerts, Secretary of State for Public Health and the Environment.

1.2.7 Ministerial Environment Committee

Within the national government there is a *Ministerial Environment Committee,* which saw the light of day with the Royal Decree of 1 March 1972. Its creation was confirmed by the Royal Decree of 30 May 1974. The Ministerial Environment Committee decides on measures concerning the fight against pollution, the quality of life and its enhancement.

1.3 CONSULTATIVE BODIES ON ENVIRONMENTAL MATTERS

Among the (all too) numerous consultative bodies, the most important, with regard to environmental matters, are the following:

1.3.1 Consultative Commission on the Environment

This was instituted in 1977 by the Minister of Public Health and the Environment. Even though it had no formal juridical basis, this Commission has functioned satisfactorily for several years, providing the Minister with opinions on various concrete problems. Its members include representatives from environmental protection associations. This national Commission will be replaced by three Regional Commissions. A 'Flemish Coucil for the Environment' was created by the Royal Decree of 16 December 1981.[11]

1.3.2 Ecology Commission

The Royal Decree of 19 March 1977 created the Ecology Commission within the Department of Public Works. This Commission is responsible for examining the ecological repercussions of all major infrastructure works proposed by the Department of Public Works, and for suggesting measures to improve locations where infrastructure works have already been carried out. The members of this Commission are independent experts.

1.3.3 Industrial Ecology Commission for the Walloon Region

This was created by Royal Decree of 16 May 1980, and comes under the Minister responsible for the economic expansion of the Walloon Region. This Commission is responsible for providing the Minister with opinions concerning the effects of industrial projects on the environment and for promoting awareness on the part of the business community with regard to the legal and regulatory provisions in force concerning environmental protection.

1.3.4 Interministerial Commission for the Co-ordination of the Prevention and Control of Atmospheric Pollution

See Chapter 3, section 5.

1.3.5 Interminsterial Water Commission

This was created by the Royal Decree of 16 May 1969, and is responsible for ensuring permanently and in general:

the co-ordination of studies on problems relating to water carried out by different ministerial departments;

the provision of opinions with the view to government intervention on the subject.

In particular, it is responsible for providing opinions on studies concerning:

the up-dating of a permanent inventory of water resources and of the annual hydrologic survey for Belgium;

the inventory of requirements for the various different qualities of water;

the harmonisation of legislation relating to water;

the co-ordination of intervention by administrations and public bodies both on a technical level and with a view to achieving regional, national or international objectives.

It is made up of:

(1) a president, appointed by the Minister of Public Health, selected from within or outside the administration;

(2) three members from the Ministry of Economic Affairs;

(3) one member from the Ministry of Agriculture;

(4) three members from the Ministry of Public Works;

(5) three members from the Ministry of Public Health;

(6) a member from the Ministry of Foreign Affairs, acting on all international aspects;

(7) a member from the Ministry of the Interior.

1.3.6 Superior Council for Water Distribution

This was created by Ministerial Decree on 3 May 1948, and is currently in the process of reorganisation.

1.3.7 Interministerial Committee on Nuclear Safety and State Security in the Nuclear Field

See 9.5.

1.3.8 Superior Council of Public Hygiene

This was instituted by Royal Decree on 15 May 1849 and reorganised by Royal Decree on 14 September 1919. It is responsible for:

(1) studying and researching everything which could make a contribution to progress in matters of hygiene, and making useful proposals on the subject;
(2) providing opinions on questions of health and hygiene put to it by the Government, on its initiative or by request from the provincial or communal authorities.

1.3.9 Consultative Commission on Foodstuffs

Article 22 of the Law of 22 January 1977 relating to the health protection of consumers with regard to foodstuffs and other products provides for the creation within the Ministry of Public Health of this Commission.

At the request of the Minister responsible for Public Health, the Commission provides an opinion on any problem relating to foodstuffs and other products covered by the law. The opinion of the Commission is required for decrees issued in implementation of the law and those concerning the standards of composition and labelling of foods and other products covered by the law, with the exception of decrees issued in implementation of international obligations and decrees for which the law requires the opinion of the Superior Council of Hygiene.

1.3.10 Permanent Committee for the Recycling of Waste Paper and Board

This Committee (known by its abbreviation 'Corepa') was set up by Royal Decree of 28 October 1976. Its objective is the encouragement and promotion, in the general interest and by all appropriate means, of development of recovery and reuse of cellulose fibres and, in particular, the development of co-operation between all parties concerned with the problem. To this end, it can:

(1) study and implement a system for the progressive increase and harmonious collection of waste paper by the various authorities concerned;

(2) provide the ministers and public bodies responsible with opinions, on its own initiative or by request, particularly on matters of financing and general policy relating to waste paper recycling;

(3) inform the public and stimulate general awareness on the subject of the fight against wasting paper;

(4) encourage the use of recycled paper;

(5) develop international contacts in the field of waste paper recovery and recycling.

1.3.11 Commission for the Scientific Study of Technical and Economic Possibilities for Alternative Methods of Agriculture and Horticulture

This Commission, instituted under a Ministerial Decree of 17 March 1977, comes within the Ministry of Agriculture. It is responsible for the co-ordinated study of the possibilities offered by alternative methods of agriculture and horticulture.

1.3.12 Consultative bodies

In addition to the above, the existence of other purely, consultative, bodies must at least be mentioned; for example:

Superior Council for Nature Conservation;

Superior Council for Forests;

Superior Council for River Fish and Fish Farming;

Superior Council for Hunting;

Royal Commission for Monuments and Sites.

1.4 NON-GOVERNMENTAL ASSOCIATIONS

The 150-odd associations concerned with the protection of nature and the environment are nearly all grouped into four federations:

Bond Beter Leefmilieu—Vlaanderen (Flemish Region);

Interenvironnement—Wallonia (Walloon Region);

Raad voor het Leefmilieu te Brussel (Flemish association for the Brussels Region);

Interenvironnement—Brussels (French-speaking association for the Brussels Region).

Their activities can be summarised as follows:[12]

(1) Constant pressure on the public powers, social speakers and all those who could be described in sociological terms as 'decision-makers', to ensure that the problems of the environment and quality of life are kept in the forefront at all levels.

(2) Co-ordination of activities by all concerned with environmental protection in order to face up to the major problems of the moment, following the conclusions arrived at by specialist working parties (e.g. nuclear questions, motorway infrastructure, leisure policy).

(3) Constant activity in the press and other media to ensure that the public is clearly and comprehensively informed on the partial and overall choices implicit in the protection of the environment and quality of life. The preparation of numerous publications.

(4) Intervention at local, regional and national levels by request from member associations, voluntary groups or individuals, on relevant questions (several hundred requests for intervention per year on such diverse subjects as the classification of a site, the protection of the population against pollution, the construction of a holiday village, the mechanical cleaning of a river, etc.).

Notes

1. Ch. Lambotte, La demembrement de la norme juridique en Belgique, *Bulletin of the IDEF,* 1980, no. 34, p. 8.
2. *Moniteur belge,* 15 August 1980.
3. Excluding questions relating to the political compromise concerning the coexistence of public and private schools, the obligation to attend school, teaching structures, diplomas, subsidies, treatments, general education standards; the exception is far more important than the rule!
4. The brief summary of the judicial structure which follows only covers the essential points; no mention is made of the Court of Assizes and the military jurisdictions.
5. J. Velu, La protection juridictionelle du particulier contre le pouvoir executif en Belgique, in *Gerichtsschutz gegen die Exekutive, Länderberichte* t. ler, Cologne, 1969, p. 74.
6. A. Mast (Précis du droit administratif Belge), 1966, no. 281, p. 210.
7. *Journal des Tribunaux,* 1971, pp. 460–474, with conclusions by W. J. Ganshof van der Meersch.
8. Compare with the Directive of the Council of the European Communities of 15 July

1975 relating to waste, which uses the term 'disposal' in a broad sense: 'the collection, sorting, transport and deposit on land' and 'necessary processing operations for their reuse, recovery or recycling'.

9. Report by Mr Cooreman and Mr Goossens, *Doc. parl.*, Senate, 1979–1980, doc. no. 434–2, pp. 103–104.

10. The distribution of responsibilities among the various national government ministries and regional executives has developed very rapidly in recent years. For this reason it seemed preferable to restrict ourselves in the following chapters to indications on the subject contained in the different legislative texts; e.g. the operating authorisation for certain classified establishments is granted, according to the General Regulation, on the proposal of the 'Minister of Labour and Employment'. In fact, these authorisations are granted on the proposal of Mr Lenssens for the Flemish Region, Mr Féaux for the Walloon Region and Mrs Neyts-Vyttebroek for the Brussels Region. We have preferred to refer only to the 'Minister of Labour and Employment' in order to simplify a text which is already all too cumbersome; the reader should refer to 1.2.6 to discover which Minister is responsible for which Region, at least when reading these lines.

11. *Moniteur belge,* 10 March 1982.

12. Information provided by Interenvironnement—Wallonia.

2

The Control Of Dangerous, Dirty Or Noxious Installations

2.1 INTRODUCTION

As has already been noted, like most other countries in western Europe, Belgium has a wide variety of legal controls and regulations covering the protection of the environment in all its different aspects. In addition to this multitude of texts, there is also a system of classified installations which has been in existence for a long time. This regulation dates back to the Imperial Decree of 15 October 1810.

The regulation currently in force is found under Title I of the General Regulation for Protection of Labour, approved by decree from the Regent on 11 February 1946. The legal basis of the general regulation is the law dated 5 May 1888 relating to the monitoring of dangerous, dirty and noxious installations, as well as steam machinery and boilers.

The purpose of controlling dangerous, dirty and noxious industries is to reconcile the private interests of these installations with those of neighbouring property owners and the general interest. In order to achieve this objective, three basic principles have been introduced into the general regulation:

(1) No installation covered by the regulation can be constructed, converted or relocated without authorisation from the responsible administrative authority (article 1, 2nd paragraph, General Regulation).
(2) Every installation covered by the regulation is subject to constant administrative monitoring (article 19, General Regulation).
(3) The authorisations granted cannot prejudice the rights of third parties (article 24, General Regulation).

The objective of the authorisation procedure is not to repress all nuisances: the authority granting the authorisation must only ensure that the neighbours do not suffer nuisance in excess of normal neighbourly conditions.[1] It is therefore very evident that the objective of classifying industries is to protect the public and not—or at least not directly—the protection of the environment as a whole.

2.2 FIELD OF APPLICATION

2.2.1

The General Regulation applies principally to 'manufacturing installations, factories, workshops, shops, warehouses, open air quarries, machinery, apparatus etc., the existence, exploitation or operation of which may cause danger, soiling or inconvenience' (article 1, paragraph 1).

Under the terms of this regulation, the field of application includes not only commercial and industrial exploitation but also machinery and equipment for private use, as exemplified in the wording of article 1(2).

Not only the exploitation, but also the very existence of the installation comes under the application of the General Regulation.[3]

2.2.2

The General Regulation does not give a specific definition of danger, soiling or inconvenience. It is left to the responsible authority to establish the nature of the installation and to decide whether or not to grant an authorisation.

In order to apply the General Regulation it is, however, not sufficient for the operation of an installation to be effectively dangerous, dirty or noxious; in addition to this, the installation must be specifically mentioned in Chapter II, Title I of the General Regulation: 'List and classification of dangerous, dirty and noxious industries'.

At the same time, the fact of being mentioned in the list is sufficient for the installation to be covered by the General Regulation, even if the installation in question does not in fact cause any inconvenience as a result of its size, location or operating conditions.

2.2.3

The list referred to above classifying all 'installations' which are dangerous, dirty or noxious in character is the object of an irrebuttable legal presumption.

In addition, the list includes an explanation (as an example) of the nature of the nuisance and indicates the services which must be consulted during the course of investigations following an application for authorisation.

The following examples illustrate this point:

Installation	Class	Type of nuisance	Services to be consulted
Manufacture of acetone	1	Great fire danger; strong and disagreeable odours	Medical inspectorate of labour
Breweries	1	Smoke; abundant release of sickly and noxious vapour	—
Bakeries and confectioners	2	Smoke, fire danger	—
Dance halls including dance floors of more than 100 m²	2	Noise, danger of fire and panic	—

The installations in each of the two classes can be subdivided into two groups under the respective supervision of the Minister of Employment and Work and the Minister of Public Health.

2.2.4

It can be seen from the wording of Title I of the General Regulation[5] that there is a special control for certain types of installations:

(1) Mines, underground quarries and open-cast mines: the co-ordinated laws of 15 September 1919;
(2) Steam apparatus: Title IV of the General Regulation (articles 724–828);
(3) Explosives factories and warehouses: Law of 28 May 1956 and Royal Decree of 23 September 1958;

(4) Nuclear reactors and installations using radioactive materials: Royal Decree of 28 February 1963 acting as a General Regulation relating to the protection of the population and workers against the dangers of ionising radiation;[6]

(5) Toxic wastes: Law of 22 July 1974 and Royal Decree of 9 February 1976 acting as a General Regulation on toxic wastes.[7]

As a result of the Decree of 8–10 July 1791, military installations do not come under the general regulation, in order to safeguard the obligation of secrecy imposed by national security.

Finally, there is a particular control for temporary installations of a duration of up to three months (twelve months for open air quarries and construction yards): investigation into the convenience or inconvenience caused and production of plans are not required; the College of Burgomaster and Aldermen is responsible for the decision (article 16, General Regulation).

2.3 OBLIGATION TO OBTAIN AN AUTHORISATION

2.3.1 Purpose of the authorisation

An authorisation is required for the construction, conversion and relocation of a classified installation (article 1(2) of the General Regulation). In addition, an authorisation is required for any extension to an installation which already holds an authorisation if it would come under a new heading of the General Regulation or if it could increase inconvenience (article 14, General Regulation).[8]

A new authorisation must be requested in the three following instances (article 15 of the General Regulation):

for an installation or part of an installation which has not commenced operations within a period established in the authorisation decree;

for an installation or part of an installation remaining inoperative for at least two consecutive years;

for an installation or part of an installation which has been temporarily put out of action for any reason.[9]

2.3.2 The decision making procedure

2.3.2.1 THE ORGANISATIONS RESPONSIBLE

Applications relating to Class 1 installations are the responsibility of the deputed states (Executive Board of the Province). Applications relating to Class 2 installations are the responsibility of the College of Burgomaster and Aldermen (article 2(1), General Regulation).

The provincial or communal authorities responsible are those in charge of the area in which the operation is carried out (article 25, General Regulation).

Applications concerning Class 1 and 2 establishments at the same time are dealt with by the deputed states. Applications concerning additions to establishments already authorised are dealt with by the authority which gave the first authorisation, except where a Class 1 addition is made to a Class 2 establishment, when the deputed states are responsible.

Nevertheless, the fact that it was the deputed states in the second instance who granted the authorisation for a Class 2 installation to operate for a specified period of time, does not, in the first instance, empower them to make the decision on an application to extend this authorisation. The existence of an administrative practice in this context is without relevance.[10]

2.3.2.2 INVESTIGATIONS PRIOR TO DECISION

Article 3 of the General Regulation indicates the elements to be included in the authorisation application, as well as the cases which require the inclusion of plans; in particular, the application must indicate the planned measures intended to prevent or reduce any nuisance in the neighbourhood which could arise from the installation.

The application[11] is presented to the responsible authority by the operator of the installation and not by the proprietor of the building and/or the installation, if this is a different person.[12]

2.3.2.2.1

A public inquiry ('enquete de commodo et incommodo') is set up by the College of Burgomaster and Aldermen.[13]

The reason for this inquiry is that the greatest possible publicity should be given to the application in order to provide an opportunity for comment from property owners and principal occupants of buildings within a specified radius, as well as public administrations, so that all the valid arguments against the proposed installation can be assessed. However, all the forms in which the investigation is to be carried out are not laid down in an obligatory fashion. An omission cannot give rise to cancellation of the authorisation if it is evident that it does not damage the interests of the persons and authorities concerned.[14]

2.3.2.2.2

The inquiry opens with the publication of the authorisation application by the posting of notices and written notification (article 4, General Regulation).

Within five days after receiving the application, the College of Burgomaster and Aldermen must post up an announcement indicating the purpose of the application. The notice must be posted both at the proposed site for the installation and on other normal hoardings as well as on hoardings in any neighbouring areas which come within 50 metres of the proposed installation, in the case of a Class 1 installation. These notices must remain on view for at least a fortnight (article 4(4), General Regulation).

The regularity of publication and proof of publication must be incontrovertible.[15]

In accordance with article 4, final paragraph, of the General Regulation, for Class 1 installations the College of Burgomaster and Aldermen must, at the same time, send individual written notification of the application to the homes of property owners and principal occupants of buildings located within a 50 metre radius of the installation as well as to the public administrations responsible for communications, water supply and other works located in the same area.

The posting up of a notice at the proposed site and on the normal hoardings does not exclude the possibility of the communal administration sending written notification of the authorisation application to individuals residing outside the proposed perimeter.[16]

2.3.2.2.3

During the course of the inquiry, all interested parties can examine the authorisation application and any plans attached to it (article 5, General Regulation).

The right of objection is not restricted to individuals who have received written notification at home of the authorisation application.[17]

25

A member of the College of Burgomaster and Aldermen or an official from the communal administration delegated to this effect[18] collects the written comments.

When this period has expired, a meeting is held during which anyone desiring to speak may be heard. At the close of this meeting, a written report of the proceedings is prepared.

The comments made either in writing or verbally during the course of the inquiry are communicated to whoever requests this information, e.g. the applicant. There is, however, no obligation to communicate automatically (*ex officio*) these comments to the applicant,[19] which undoubtedly constitutes a shortcoming in the regulation in force.[20]

With regard to Class 1 installations the communal administration sends the governor of the province the complete file, including the opinion of the College of Burgomaster and Aldermen, within ten days from the close of the inquiry.

2.3.2.2.4

After the close of the inquiry, the opinions of a certain number of technical officials—indicated in articles 7 and 8 of the General Regulation[21]—and of the provincial town planning director are requested (article 9, *bis*, General Regulation). These opinions constitute a substantial formality.[22]

The responsible authority can, in addition, consult any authority it may consider useful without giving any reasons for doing so.[23]

2.3.2.2.5

Neither the comments given during the inquiry nor the technical opinions have an obligatory character. The authorities granting the authorisation take the final decision at their discretion.

2.3.3 The decision

2.3.3.1 TIME LIMIT

The responsible authority must announce its decision within a period of three months from the day when it was officially informed (article 10(1), General Regulation).

If during this time no announcement is made, the authority responsible in the second instance (the deputed states for Class 2 installations, the

King for Class 1 installations, see 2.3.4.1.1) can call in the application and take the final decision on it as a first and last resort (article 10(2), General Regulation).

2.3.3.2 OBLIGATION TO GIVE REASONS

The decision of the authority responsible for granting the installation is not a legal decision; it is an administrative decision on the subject which does not have the authority of a judgement.[24]

However, in accordance with article 10 of the General Regulation, the authority which takes a decision on the authorisation application must give reasons for the decision.

In order to conform with the obligation to justify the decision to grant an authorisation, the administrative authority must reveal the reasons which justified the granting of the authorisation. The general statement that 'the observation of the regulatory requirements and conditions is adequate to avoid the dangers and inconvenience inherent in the operation of the above mentioned installation' cannot constitute a valid justification unless it is accompanied by an explanation and discussion of the concrete elements involved, in order to prove that the authority in charge has itself also taken these elements into consideration.[25]

The obligation to give reasons can be satisfied either by the indication of the reasons which, in the opinion of the authority, justify its decision, or by reference to other opinions given, if these give reasons to justify the granting or refusal of the authorisation,[26] provided that the administrative authority has acted on these opinions.[27]

If the authority does not act on the opinions to which it refers, it must give its reasons for not doing so.[28]

2.3.3.3 POWERS TO EXAMINE OBLIGATIONS

The authority granting the authorisation has the final decision.[29] It is for this reason that in the case of Class 1 installations the deputed states are not bound by the opinion of the College of Burgomaster and Aldermen;[30] neither do the opinions of technical officials have any obligatory value.

The authority granting the authorisation must be sure that the neighbours do not suffer any inconvenience in excess of what would normally be expected, but it cannot require the installation to present no inconvenience at all.[31] In fact, it has been considered that the refusal of an operating authorisation which stated only that operation of the under-taking would result in neighbourhood inconvenience and odours could not be completely eliminated, would not satisfy the requirement for justification. It would mean, in effect, that another establishment of the same type could only be authorised if the nuisance known to all was completely eliminated, while in reality it should be investigated to determine whether the nuisances from that establishment exceeded normal neighbourly inconvenience in the area taking into account all the circumstances and available information, and whether it would be possible to reduce the nuisance to normal levels for the area by imposing operating conditions.[32]

2.3.3.4 CONDITIONS TO BE MET

The authority verifies that the following conditions are met by the operation in question:[33]

(1) the conditions required for public safety, cleanliness and convenience;
(2) the conditions required to safeguard the safety, hygiene and comfort of the workforce;
(3) the conditions required to safeguard the immediate environment.

2.3.3.5 CONSIDERATION OF PLANNING REGULATIONS

The authority taking a decision on an authorisation application must take into consideration the requirements of the planning regulations in force at the time of the decision,[34] even if it is taking a decision on appeal.[35] But at the same time, it is not bound by proposed planning regulations.

In the absence of imperative stipulations prohibiting the use of buildings for industrial purposes, it is within the assessment power of the authority in charge to decide whether a noxious business is compatible with the character of the area in which it is located.[37]

2.3.3.6 OTHER CONSIDERATIONS

The authority cannot invoke other considerations.[38] It is with good reason that H. Lenaerts states that: 'Economic considerations such as extension of the industry or motives of competition cannot be taken into consideration'.[39] The financial costs incurred by the operator as a result of the conditions imposed in the authorisation are not in themselves sufficient to justify dispensing with them.[40] In principle, however, financial costs which would render the operation economically unviable cannot be imposed on a business.[41]

2.3.3.7 CONDITIONS

The responsible authorities can only decide on an application in the form in which it is submitted to them.[42] However, they can grant only a partial authorisation, that is, for only a part of the application. When they grant an authorisation, they determine the conditions with which the operator must comply.[43]

These conditions which control the operation must be laid down in a sufficiently clear manner. In fact, the authorisation is a unilateral act on the part of the authority which must clearly limit the extent of that authorisation.[44]

The conditions imposed vary according to the case: restrictive measures, the fixing of emission standards, etc. However, the conditions imposed must be achievable by the operator himself.

2.3.3.8 DURATION OF VALIDITY

Authorisations cannot be granted for a period of more than thirty years. They can be renewed (article 11(2), General Regulation). The fixing of a period of validity for the authorisation under article 11 necessarily supposes that when the term expires, the authority responsible for investigating the renewal application will examine any new elements which may have arisen concerning the operation itself, and also any other questions such as the current requirements of the location in relation to approved town planning policy, etc.[46]

An authorisation can be granted for a trial period of up to two years (article 11(3), General Regulation).

The authorisation fixes a period of time within which the installation must commence operations. This period cannot exceed two years (article 11(1), General Regulation).

2.3.3.9 PUBLICITY CONCERNING THE DECISION

The decision is made public by written notification and posted notices (article 12, General Regulation).

A copy of the decision is sent to the applicant, to the technical official in charge of monitoring and, when the decision does not come from them, to the College of Burgomaster and Aldermen, and to the public administrators to which a notification relating to the inquiry had to be sent.

Within five days from receipt of the decision by the communal administration, it must be made public by posting notices outside the town hall and at the site of the installation. These notices must remain on view for a period of 10 days. The notification of the decision must not precede the posting up of notices.[47]

The obligations imposed regarding the publication of the decision are designed to ensure the safeguarding of the rights of interested parties by the widest possible publicity. In no case can the posting up of notices operate in such a way as to mislead the neighbours.

2.3.4 Methods of appeal

2.3.4.1 ADMINISTRATIVE APPEAL

Article 13 of the General Regulation provides for an administrative appeal. All interested parties (in the broadest sense—the property owners and occupants of neighbouring plots, including those to whom a written notification should be sent; the public powers; an association for the protection of the environment; briefly, anyone who might suffer inconvenience as a result of the operation) can appeal to the deputed states against the decision of the College of Burgomaster and Aldermen.

An appeal can be made to the King against decisions in the first instance by the deputed states, by the governor of the province acting on their own initiative or at the request of a technical official or by the communal authority or any other interested party.

The appeal should be sent by registered letter post within ten days from the first day that the decision notices appeared; this time limit is compulsory (article 13(3), General Regulation).[43]

The appeal has no suspensive effect: the operation of the installation can be started or remains prohibited, depending on whether the authorisation is granted or refused.

2.3.4.1.1

This form of appeal is an administrative appeal. 'Administrative appeal has the advantage over legal appeal of allowing the party not only to denounce the illegality of the measure by which it is affected, but also to enable an assessment to be made of its inopportuneness or ineffectiveness or even to appeal to the goodwill of the administration'.[49]

The resolution taken as a result of an appeal of this kind is not a judicial decision and does not have the authority of a judgement.[50]

The authority who takes a decision on appeal has the same responsibilities as the authority who made the decision in the first instance. It can order a new inquiry but is not obliged to do so.[51] A new consultation with the Provincial Town Planning Director is, however, obligatory.

Considering the character of the appeal, the applicant has no other guarantee in defending his interests than the laws and regulations which specifically apply to his case. It is in this way that the applicant need not be heard;[52] nor does the decision have to be pronounced publicly, since article 97 of the Constitution is only applicable to the courts and tribunals.[53]

2.3.4.1.2

The appeal and the subsequent decision are notified to the communal administration by the governor of the province; they are brought to the notice of interested parties by the posting up of notices and by written notification according to the method and timetable laid down by article 12 of the General Regulation (see 2.3.3.9).

Irregularity in the notification procedure does not adversely affect the legality of the decision.[54]

2.3.4.2 APPEAL TO THE CONTROLLING AUTHORITY

The provincial and local authorities are subject to what could be called 'administrative control'. All their decisions can be suspended or cancelled by the Governor or the King if they are contrary to the law or the general interest.[55]

Any interested party can therefore request the Governor or the King, depending on the case, to suspend or cancel a decision referring to an authorisation application. The protection authorities are not, however, obliged to take any decision, whatever it would be, nor to take the request into account.

This possibility can be of interest when the period of time established for an administrative appeal has expired.[56]

2.3.4.3 JURISDICTIONAL APPEAL

2.3.4.3.1

After exhausting all the administrative appeals under article 13 of the General Regulation, any interested party can address an appeal to the Council of State; this appeal is only admissible if the applicant can prove injury to a direct, actual and legitimate personal interest.

> The general and impersonal interest every citizen has in the correct enforcement of the law is not sufficient. An identifiable relationship must exist between the applicant and the incriminating decision. . . . The interest must exist right up until the moment when the Council of State makes its decision. . . .

Recent decisions taken by the Council of State admit that associations, the social aim of which is to monitor the protection of the environment, have a sufficient legal interest to appeal for the cancellation of decisions granting authorisation. The high courts made the following statement (*Bond Beter Leefmilieu* versus *Minister of Public Works:*)

> Given that the by-laws of the plaintiff association describe the aim of the association as being the protection of the integrity and variety of the environment and the achievement of a suitable quality of life, an object which notably leads to co-ordination and support of efforts which are pursued in this respect by private institutions fully described in the statutes, this association can intervene in the defence of the environment by means of an independently introduced request to the Council of State.

> Since 'Bond Beter Leefmilieu' is a 'head' association made up of 35 associations, each of which on its own accord is engaged in the environment and which are distributed throughout the country, the association is undoubtedly representative in environmental protection matters. Because of this, the association draws the legal interest required, and in particular the specific group interest in the cancellation of a building permit in as much as it disturbs the environment. It is not demonstrated that the association in question acts only in the defence of the individual interests of its members.[58]

The Council of State examines on its own initiative whether or not the appeal can be received.[59] The appeal has no suspensive effect.

2.3.4.3.2

The Council of State can examine only the question of the legality of the decision in the second instance.

Contrary to the procedure under French legislation, the Council of State in Belgium is not able to take the place of the administrative authorities from the point of view of assessing the specific circumstances involved; the authority granting the authorisation has the final decision on the dangerous, dirty or noxious character of the installation.[60]

It should, however, be underlined that the control of whether the decision is based on proper considerations constitutes a control of legality: 'In examining the motives, the Council of State does not consider the opportuneness of the decision submitted to it for censure; it only examines it with reference to the regularity of the act'.[61]

2.3.4.4 APPEAL TO THE COURTS

The applicant can claim damages from the civil judge if the authority has wrongfully intervened, for example in the case when authorisation has been refused in an irregular manner or has only been granted after an unjustifiable delay.[62]

Equally, the operator of an installation who is the object of penal proceedings or of a civil trial, brought about by an injured third party, can always invoke the *illegality* of the decision, if he considers that the authorities have taken an unlawful decision in his regard.

2.4 MONITORING AND IMPLEMENTATION OF THE AUTHORISATION

2.4.1 Monitoring

2.4.1.1

The operating authorisation is valid from the moment that the installation has completely satisfied all the general regulatory requirements and special conditions imposed by the authorisation decree.[63]

Monitoring is carried out by the technical officials listed in article 7[64] and the Burgomaster (article 20, General Regulation).

The technical officials have free access to installations. They carry out the overall supervision of the installation, whereas the duties of the Burgomaster are restricted to the following:

(a) He must check that installations engaged in operations are in possession of a valid authorisation.
(b) He must ensure that the operating conditions imposed by the College of Burgomaster and Aldermen[65] are observed.

The supervision and control of the conditions imposed by the deputed states are the exclusive responsibility of the technical official.

2.4.1.2

If a danger jeopardises the safety and health of the personnel or neighbours and the director of the business refuses to carry out the instructions of the responsible technical officials, the Burgomaster, on the receipt of a report from the technical officials in charge, can order the cessation of work or can seal off the machinery or, if necessary, can proceed with the provisional closure of the installation, depending on the character of the danger (article 21, General Regulation).

The technical official holds a similar responsibility in the case of inaction on the part of the Burgomaster or if the danger is so imminent that any delay could result in an accident.

In either case, the operator can make an administrative appeal to the King. This appeal has no suspending effect.

2.4.2 Sanctions

2.4.2.1 ADMINISTRATIVE SANCTIONS

2.4.2.1.1

According to article 19 of the General Regulation, the authority which has granted the authorisation can take two types of measure: on the one hand, if the director of the business does not observe the requirements or conditions imposed under the authorisation, the authority can withdraw or suspend the authorisation; on the other hand, it can impose new conditions on the installation.[66]

Administrative appeal against these administrative sanctions has the effect of suspending the decision in question.

No modification in the operation or control to which an installation is subject is required before new conditions can be imposed.[67]

In fact, with reference to noxious installations under the General Regulation on the Protection of Labour, acquired rights cannot be invoked based exclusively on the consideration that the installation has been in operation for a long time.[68]

2.4.2.1.2

In addition, the Burgomaster and the technical official responsible can take the measures listed in article 21 (see 2.4.1.2):

(a) In the case of the erection, conversion or relocation of an installation without authorisation when authorisation is required; these measures are lifted when the necessary authorisation is granted;
(b) When the conditions controlling the operation are not observed. In the case when the authorisation has been granted by the deputed states or by the King, their preliminary approval is required.

2.4.2.2 PENAL SANCTIONS

Contraventions of the regulations on dangerous, dirty or noxious installations can be dealt with by corrective punishment (Law of 5 May 1888).

The police court is not empowered to order the closure of an installation or to remove an authorisation; only the administrative authority can take such action.[69] However, the courts can order the closure of an installation if it has been operating without authorisation from the administration.

2.4.2.3 CIVIL RESPONSIBILITY

In accordance with article 24 of the General Regulation, the granting of an authorisation does not prejudice the rights of third parties.

Civil law provides a compensation obligation in two cases:

In the case of fault on the part of the person causing the damage; under article 1382 of the Civil Code, all damage resulting from fault must be compensated;

In the case where there is no fault but the problem relates to abnormal disturbance of the neighbourhood; since the neighbouring property owners all have an equal right to enjoy their property, once the relationship between the properties involved is established, and bearing in mind normal neighbourhood conditions, a balance is established between the respective rights of the property owners. The owner of a building who, through no fault of his own, disrupts this balance by subjecting another property owner to disturbance exceeding normal neighbourly nuisance, is liable to pay just and adequate compensation to the injured neighbour in order to reinstate that balance.

The action based on article 544 of the Civil Code (the theory of neighbourhood disturbance which permits compensation without fault, is subsidiary to article 1382 of the Civil Code (liability in tort). The distinction between the two actions is of great practical importance: the damage caused by an error (fault) must be completely repaired; damage caused without fault must only be repaired to the extent to which such damage is abnormal.

Notes

1. Council of State, 3 December 1976, no. 17.976, Brasseur; 12 October 1976, no. 17.818, Cornelis; 19 April 1971, no. 14.675, Mollers.
2. Council of State, 14 May 1974, no. 16.412, Paque.
3. H. Lenaerts, Gevaarlijke, ongezonde en hinderlijke inrichtingen, in *Administratief Lexicon* 1962, no. 10; M. A. Flamme, Les establissements dangereux face a l'administration et aux voisins, in *L'entreprise dans la cité,* Liege/La Haye, 1967, p. 233. According to J. M. Favresse, more subtlety is required: the simple existence of an installation would not make the regulation apply except where the very nature of the plant means that it is classified because of its very existence; J. M. Favresse, Etablissements dangereux, insalubres ou incommodes, in *Repertoire du droit de la construction,* 1970, p. 9.
4. H. Lenaerts, *op.cit.* no. 9.
5. Heading I—control of classified installations as dangerous, dirty and noxious, with the exception of mines, mining and underground quarries, as well as explosives factories and storage depots.
6. See Chapter 9.
7. See Chapter 8.
8. It is only in this case that the authorities decide on the usefulness of an inquiry, which in principle is compulsory. Council of State, 11 April 1970, nos 16.969 and 16.970, Cornières R. Chartier and Son and J. Lux: 'The regulations of article 14 of the General Regulation on the one hand states that an extension to an authorised installation is not in itself "of a nature to increase the danger, dirtiness and inconvenience inherent in the operation" and on the other hand that even if the extension is of a nature to lead to such an increase, the authority can still assess whether or not a new inquiry is necessary, taking into account the particular circumstances involved.'
9. For an actual case, see Council of State, 18 November 1955, no. 4670, Van Lierde.
10. Council of State, 22 May 1956, no. 5141, Commune of Hoboken.
11. For applications relating to Class 1 installations a special formula is envisaged (Ministerial Decree of 11 February 1947).
12. Cass. 7 January 1952, *Pas.* 1952, I. 227.
13. Applications relating to Class 1 installations addressed to the deputed states are sent back to the communal administration within a period of two days after receipt (obligatory period)—to permit the inquiry to be organised.
14. Council of State, 7 July 1954, no. 3.556, Gruber.
15. Council of State, 6 January 1976, no. 13.755, Andries and cons.
16. Council of State, 13 October 1959, no. 7.282, Hollaert.
17. See note 16.

18. In fact, the police commissioner or secretary of the commune.
19. Council of State, 4 October 1960, no. 8107, Wimme.
20. H. Lenaerts, *op. cit.* no. 45.
21. The engineers and technical work organisation, the officials of the hygiene administration, the engineers of the mining administration, officials of the explosives service, technical officials in charge of monitoring by the Minister of Public Health.
22. Council of State, 25 March 1958, no. 6192, Ongenae.
23. Council of State, 14 May 1957, no. 5645, Van Doorslaer.
24. Council of State, 14 July 1955, no. 4508, Maes and cons.; 4 October 1960, no. 8107, Wimme; 22 January 1963, no. 9827, Moerman; 26 November 1963, no. 10.292, Hutsebaut; 20 December 1963, no. 10.349, Vandevijver; 17 February 1976, no. 17.439, Peeters and cons.; 30 April 1976, no. 17.594, Kuzniak.
25. Council of State, 15 April 1977, no. 18.225, Commune of Braine-l'Alleud.
26. Council of State, 19 February 1974, no. 16.253, Knaepen; Council of State, 13 April 1976, no. 17.564, s.p.r.l. Cecima; 20 February 1979, no. 19.457, s.p.r.l. J. Lemaire and son.
27. Council of State, 6 October 1978, no. 19.175, Delanois; 29 March 1977, no. 18.205, s.p.r.l. Intergarde.
28. Council of State, 4 May 1976, no. 17.595, Lorman and cons; 3 December 1976, no. 17.975, Navette; 6 February 1979, no. 19.426, De K ck.
29. Council of State, 20 June 1952, no. 1661, Commune of Aywaille; 13 October 1959, no. 7282, Hollaert.
30. Council of State, 24 January 1958, no. 13.016, Mines of Limbourg/Meuse.
31. Council of State, 3 December 1976, no. 17.976, Brasseur.
32. Council of State, 31 January 1978, no. 18.727, Lavent.
33. Council of State, 11 June 1968, no. 13.016, Mines of Libourg/Meuse.
34. Council of State, 26 April 1955, no. 4238, Hoboken Commune; 27 March 1959, no. 6996, Ouchakov-Renard; 30 September 1960, no. 8082, Vanseer; 26 February 1972, no. 16.265, Debedts. The authority receiving an application for an extension to an installation already authorised by a previous decision must take into account town planning regulations in force at the moment when a decision is taken on the extension application: Council of State, 20 April 1979, no. 19.580, Gillet.
35. Council of State, 8 November 1977, no. 18.527, Commune of Brasschaat.
36. Council of State, 5 March 1957 and 24 January 1958, nos. 5533 and 5994, Hoboken Commune; 17 June 1960, no. 7935, Meeus; 30 November 1962, no. 9742, Hockers; 10 December 1976, no. 17.992, Forest Commune.
37. Council of State, 10 December 1953, no. 2980, Caluwaerts; 13 July 1956, no. 5273, Cuypers; 4 April 1958, no. 6219, Brunson; 17 November 1961, no. 8965, Ronge and the Commune of Ixelles; 28 April 1966, no. 11.776, Town of Courtrai. See also Council of State, 13 February 1979, no. 19.432, De Win: 'It cannot be purely and simply deduced from article 5.1.0 of the Royal Decree of 28 December 1972 that noxious installations can in any case operate if the inconvenience caused can be reduced to a level compatible with the use of the habitat zone by the imposition of operating conditions, among others, in as much as they can be considered to apply to the normal equipment of a residential area and to be compatible with the immediate environment.'
38. See *Repertoire Practique de Droit Belge V*, 'Establissements dangereux, insalubres et incommodes', nos. 51–52 and legislation cited; Ch. Goossens, Le Regime Juridique des Autorisations requises pour l'Exploitation des Etablissements Industriels en Belgique, *Revue Internationale des Sciences Administratives*, 1953, pp. 630–631.
39. H. Lenaerts, *op. cit.* no. 51.
40. Council of State, 7 October 1960, no. 8118, Central Construction Company s.a.
41. Council of State, 10 April 1965, no. 10.548, Colin and Petitpas; see also the conclusions of the Procurer General Leclercq, Cass. 25 May 1950, *Pas.*, 1951, I, 7.
42. Council of State, 22 December 1959, no. 7513, Uvel.

43. Council of State, 9 December 1949, no. 175, Commune of Jupille.
44. See decree mentioned in note 41, Colin and Petitpas.
45. Council of State, 11 April 1975, nos. 16.969 and 16.970, Carrières R. Chartier and Son and J. Lux.
46. Council of State, 26 October 1956, no. 5348, S. A. Carrosserie Paul d'Heure; 4 October 1960, no. 8107, Wimme.
48. The period in which the communal administration must proceed with the posting up of notices of the decision and the period during which the interested parties can appeal are, of course, quite different in character: the former is a fixed time allowed by order and the latter is an obligatory period of time. The posting up of notices of the decision, even if it does not occur within the period envisaged, gives rise to the period in which appeal is made (Council of State, 21 January 1977, no. 18.057, Van Deyck et cons).
49. Andre Mast, Précis de droit administratif Belge, 1966, p. 307.
50. Council of State, 20 December 1963, no. 10.349, Vandevijven; cf. P. Vermeulen, Establissements dangereux, insalubres ou incommodes. Demandes d'autorisation en vue de leur exploitation. Decision de la deputation permanente statuant en appel. Nature de ces decisions. Revue Administrative, 1953, 53.
51. Council of State, 14 May 1957, no. 5645, Van Doorslaer; 5 October 1965, no. 11.228, d'Hondt.
52. Council of State, 26 November 1963, no. 10.292, Hutsebaut.
53. Council of State, Van Doorslaer, mentioned in note 51.
54. Council of State, 7 July 1957, no. 3556, Gruber; 23 February 1960, no. 7646, Baetsle; 26 November 1965, no. 11.531, Ets. René Joassim, s.p.r.l; 18 March 1966, no. 11.707, Imprimerie des Editeurs s.a.
55. Article 108 of the Constitution, articles 89 and 125 of the Provincial Law, articles 86 and 87 of the Communal Law.
56. M. A. Flamme, op. cit. no. 10.
57. A. Mast, op. cit. no. 523.
58. Council of State, 9 January 1980, no. 20.021, Bond Beter Leefmilieu-Interenvironment.
59. Council of State, 30 September 1960, no. 8081, Caes; 19 September 1961, no. 8.780, de Henau.
60. Constant jurisprudence of the Council of State.
61. A. Mast, op. cit. no. 503, p. 373, with reference to Auby and Drago, Treaty of administration contention, Heading II, no. 1163.
62. Council of State, 9 October 1964, no. 10.825, Houvenaghel: 'The tribunals alone are responsible for understanding an application for compensation based on faulty measures taken by engineers in charge of the investigation of authorisations under the general regulation.'
63. With regard to Class 1 installations (not temporary installations), article 18 of the General Regulation states that the operator is obliged to inform the technical official in charge, at least 15 days in advance, of the date fixed for the starting up of the installation.
64. See note 21.
65. The Burgomaster can, in addition, be in charge of the supervision and control, under the Royal Decree, of compliance with the operating conditions by Class 1 installations.
66. Council of State, 27 February 1964, no. 16.272, Robin and cons.
67. Council of State, 7 October 1960, no. 8118, s.a. Central Construction Company.
68. Comp. Council of State, 27 October 1976, no. 17.865; Rubrecht: 'The circumstance that people build in the surrounding area does not mean that they do not or should not accept some form of operation if it is an already existing business'; see also Council of State, 1 July 1980, no. 20.491, Devragas.
69. Court of Cassation, 27 April 1962, Pas., 1962, I, 938; see J. M. Favresse, op. cit.,

pp. 55–57; H. Cousy, De betekenis van de overheidsvergunning voor de privaatre-chtelijke aansprakelijkheid in Belgie, Zwolle, 1975, pp. 38–53.
70. Cor. Tongres, 16 January 1969, *R.W.,* 1969–1970, 87.

3

Atmospheric Pollution

3.1 THE LAW OF 28 DECEMBER 1964 RELATING TO AIR POLLUTION CONTROL

3.1.1 Introduction

In Belgium, 'where industrialisation and the increase in the number of cars on the road have assumed spectacular proportions and where domestic heating by oil has progressively replaced coal',[1] the atmospheric pollution situation became alarming by the 1960s. The promulgation of the Law of 28 December 1964 relating to air pollution control constituted the beginnings of a comprehensive policy on the subject.

3.1.2 Application

The Law of 28 December 1964 is above all characterised by its very wide field of application; pollution of the atmosphere is taken to mean 'all emissions into the air, from any source, of gaseous, liquid or solid substances which could result in harm to human health, animals or plants, or could cause damage to goods or locations' (article 2).

3.1.3 Powers

The King is empowered to take 'all appropriate measures to prevent or combat pollution of the atmosphere'.

40

In particular, he can:

(1) forbid certain specific forms of pollution;

(2) control or prohibit the use of equipment or devices which may cause pollution;

(3) impose or regulate the use of equipment or devices intended to prevent or combat pollution.

The King can also impose special conditions concerning professional qualifications and access to the profession of people engaged in the installation of equipment or devices which could affect atmospheric pollution.

3.1.4 Execution of the law

The royal decrees executing the law must be submitted to the Superior Council of Public Health for its opinion. These decrees are jointly proposed by the Public Health Minister and the minister responsible for the origin of the pollution to be controlled:[2]

by the Minister for Economic Affairs if it is a question of mines, mining or underground quarries;

by the Minister of Employment and Work for industrial or commercial installations;

by the Minister of Public Works for buildings under his control;

by the Minister of Communications for transport by road, rail or air.

The Minister of Public Health is the only person responsible for all cases of pollution which, by reason of their origin, do not come under the responsibility of any ministerial departments mentioned in the law.

He also has a co-ordinating function, which is carried out with particular reference to the following sectors: sampling and analysis of suspected polluting substances emitted into the air; research relating to the effects of atmospheric pollution; research for effective ways of combating atmospheric pollution; informing the public.

3.1.5 Interministerial Commission for the Co-ordination of the Prevention and Control of Atmospheric Pollution

The central law of 28 December 1964 has over the years been 'rendered sterile'[3] because of the lack of the necessary decrees for its execution; this was predictable, if not inevitable, since the problems created by atmospheric pollution involve several different ministerial departments. A co-ordinating action was therefore essential. It is for this purpose that the Royal Decree of 7 October 1971 set up an 'Interministerial Commission for the Co-ordination of the Prevention and Control of Atmospheric Pollution'.[4]

This Commission, composed of officials from all the ministerial departments directly or indirectly concerned with the problems created by atmospheric pollution—ten departments in all—is in charge of the following:

(1) preparing regulatory measures to be taken within the framework of the Law of 28 December 1964;

(2) giving its opinion on questions posed by the Minister of Public Health relating to the prevention of atmospheric pollution;

(3) formulating and presenting the Ministry of Public Health with all proposals considered appropriate by the Commission on the prevention of atmospheric pollution;

(4) ensuring the co-ordination of all works carried out in this sector by the various ministerial departments.

The members of the commission were nominated by the ministerial decrees of 7 September and 18 November 1975, 18 and 31 May 1976 and 23 March 1978.[5]

3.1.6 Supervision of the application of the law

To facilitate the identification of pollution sources, the Law of 28 December 1964 provides for the designation of special agents to supervise the application of the law and its executive decrees.

However, to date no royal decree has yet been promulgated designating agents with general responsibility; at most a number of royal decrees concerning certain well defined forms of pollution contain a provision for agents; they only have responsibility for matters covered by these decrees.

Be that as it may, these agents—in addition to every other officer of the judicial police—have the right to enter, day or night, all installations where they have reason to believe that illegal atmospheric pollution is being caused, with the exception of dwellings.

If sufficient indication exists to suggest that there is atmospheric pollution originating from a dwelling, it can be visited between 5.00 am and 9.00 pm by two of these agents, after obtaining an authorisation to do so from the police court judge.

In addition, these special agents are empowered to:

(1) take or have taken samples of the substances emitted into the atmosphere, as well as of materials which are presumed to be the cause of the atmospheric pollution, and to have them analysed by a suitably equipped laboratory;
(2) proceed with or cause action to be taken by suitably equipped organisations on the testing of apparatus or devices which may cause pollution or which are intended to reduce pollution.[6]

Furthermore, the agents can, where there is a risk of continuous pollution, provisionally forbid the use of apparatus or devices which, as a result of their constitution or properties, are not in a fit state to operate in a manner conforming to the executive decrees of the law, they can have them sealed off and can take any emergency measures which may be required in the circumstances, in the interests of the population and clean air.

These measures cease to have effect after a period of eight days if, during the course of that period, they are not ratified—the operators having been heard or called—by the director of the administration to which the agent who has taken the measures belongs.

Ratification decisions are notified, without delay, by registered letter to the operators of the apparatus and devices.

A non-suspensive appeal to the King against ratification decisions is open to all interested parties.

Finally, the specially designated agents can request the assistance of the communal authorities in carrying out their duties; in particular, they can request these authorities to prescribe any necessary emergency measures to deal with a serious case of actual or imminent atmospheric pollution in the interest of public health and clean air. In this case, the minister in charge of public health and the governor of the province must be informed immediately.

It should be noted that the intervention of the communal authorities can only be requested—the text makes this clear—in the case of *serious* pollution; the designated agents themselves can only act when there is a risk of *serious* damage to the population. However, the authorities can also act in the case of an emission which has already taken place and in the case of current emissions which could constitute an *imminent* act of pollution.

3.1.7 Sanctions

Article 10 of the law punishes the following acts by imprisonment of from eight days to six months and a fine of from 26 francs to 5,000 francs,[7] or by only one of these punishments:

(1) the possession of stationary or mobile articles which, as a result of negligence or lack of forethought, give rise to a form of pollution forbidden by the King;
(2) the contravention of the requirements of the royal decrees executing this law;
(3) the refusal or opposition to inspections, sampling or emergency measures provided for under the law.

These punishments can be doubled if a new contravention is committed within two years of a previous conviction for a contravention under this article.

3.2 SPECIALLY PROTECTED ZONES

3.2.1

As a result of the Royal Decree of 26 July 1971,[8] specially protected zones have been created 'on the basis of the level of pollutants recorded by the Department of Public Health and taking into consideration demographic and topographic factors'. There are at present five specially protected zones, as follows:

the Antwerp region;

the Brussels region;

the Charleroi region;

the Ghent region;

the Liège region.

These areas which have been established in particular on the basis of having an annual average level of sulphur oxides in excess of 150 micrograms per cubic metre, are subject to special measures to control atmospheric pollution from installations for space heating by combustion.

3.2.2

In the specially protected zones, the use of fuels for central heating is controlled as follows:

(1) The burning of peat, lignite, non-smokeless agglomerates, as well as wastes of any kind, is forbidden.
(2) The sulphur content of liquid fuels must not exceed 1% in weight, whatever the type of liquid fuel used.
(3) Solid fuels must not have a content of volatile sulphur exceeding 1% in weight.

The destruction by burning in the open air of wastes of any kind is forbidden, with the exception of vegetable wastes arising from garden maintenance, deforestation, land clearance or professional agricultural activities.[9]

The exhaust ducts for combustion products from heating equipment must be constructed in such a way as not to lead to localised pollution which could result in a risk to health or nuisance to individuals.

Before the entry into force of the Royal Decree of 6 January 1978 (see 3.3.1.2) in specially protected zones, central heating installations were subject to a series of special operating conditions; nowadays, the Decree introduced regulations which apply throughout the national territory.

The technical officials and agents responsible for inspection can make use of the public hygiene and nuisance inspection services under the Minister of Public Health, as well as the technical officials of the Institute of Hygiene and Epidemiology. These are the same agents specially desig-nated to supervise the application of the regulations of the Royal Decree of 26 July 1971.

3.2.3

It should be pointed out here that the regulations under the Royal Decree of 8 August 1975 relating to the pollution of the atmosphere by sulphur oxides and dust arising from industrial combustion installations are more or less severe, according to whether or not the installation is situated in a specially protected zone (see 3.3.2).

3.2.4

In addition, it should be noted that a Royal Decree of 31 March 1978 modified by a Royal Decree of 28 August 1981[10] has fixed the permissible limits for the concentration of lead in the air in the area of the Commune of Hoboken, where one of the largest metallurgical factories is situated. The annual average must not exceed:

from 1 June 1981   2.3 $\mu g/m^3$

from 1 June 1982   2.1 $\mu g/m^3$

3.3 MEASURES APPLICABLE THROUGHOUT THE COUNTRY

3.3.1 Central heating installations

3.3.1.1

The regulations of the Royal Decree of 26 March 1971 relating to the prevention of atmospheric pollution arising from combustion installations[11] apply:

(1) to combustion apparatus and equipment for space heating;
(2) to all other combustion installations with the exception of those which are required for the operation of classified installations on the list attached to the General Regulation for the Protection of Labour or the operation of other installations indicated in the list (see Chapter 2).

The Minister of Public Health, in agreement with the Minister of Economic Affairs and the Minister of Employment and Work, establishes the technical specifications to which these combustion apparatuses must comply in order to be manufactured, imported or sold in Belgium.

Once these technical specifications are in force, such combustion apparatus can only be sold if it is accompanied by a notice containing the conditions for installation and control, operation and maintenance instructions for the apparatus, as well as an indication of the fuels for which the apparatus was designed.

The agents specially designated to monitor the application of the regulations of this royal decree are the same as those who monitor the application of the regulations of the Royal Decree of 26 July 1971 on the specially protected zones (see 3.2.2).

3.3.1.2

The Royal Decree of 6 January 1978 for the prevention of atmospheric pollution arising from space heating with liquid or solid fuels[12] applies to space heating installations which come under the sole charge of the Minister of Public Health.

This decree contains obligations for users as well as for technicians in charge of the maintenance of the installations.

The *users* of heating installations must:

(1) only use fuels for which the installation is designed and regulated;
(2) maintain the installation in a proper functioning condition;
(3) arrange for maintenance of the installation—the interval between two maintenance services must not exceed fifteen months.[14]

In the event of a control test[15] not conforming to the proper functioning conditions required, the user has up to three months in which to bring the installation into good working order and to obtain proof that this has been done.

The carrying out of the annual maintenance service must be attested by a verification certificate given to the user by the technician or firm in charge of the maintenance service.

The qualifications of these technicians must be recognised by the Minister of Public Health.[16] To this effect, they must possess an aptitude and a permanent qualification certificate[17] and be in possession of the properly maintained materials necessary for the control tests. The certificate is granted for a period of five years.

3.3.1.3

The Royal Decree of 6 January 1978 also contains regulations relating to the admission regulations for new installations, i.e.:

installations brought into service for the first time after the entry into the force of the Decree;

installations where the boiler had been replaced since the entry into force of the Decree;

installations converted for the use of another type of fuel since the entry into force of the Decree.

These new installations are subject to admission verifications to be carried out by an appropriate organisation,[18] on the initiative of the person in charge of the work, before the installation is brought into service.

The admission verification includes an examination of the generator, particularly of the burner and the furnace, examination of the exhaust conduits for combustion gases and the ventilation ducts in the room where the generator is situated, an estimation of the risk of the formation of fumes, bearing in mind the characteristics of the furnace, the burner and the fuel used, the tests required to ensure the correct functioning of the apparatus. This gives rise to a detailed report which is sent to the person in charge of the work as well as to the proprietor. This report contains the result of the calculations made, the conclusions of the various verifications carried out and, if necessary, suggestions for improving the installation.

This document should be sent to the user who must produce it on request by the agents in charge of supervising the application of this Decree.

If the report on the admission verification concludes that it is necessary to modify the installation, it must be modified within a period of three months from the date of the report and examined again by the same organisation which prepared the previous report.

3.3.1.4

Monitoring compliance with the regulations in the Royal Decree of 6 January 1978 is the responsibility of the officers of the judicial police, as well as the officials and technical agents of the Ministry of Public Health, of the provinces, the associations and federations of communes and the communes to be designated for this purpose by the King.

3.3.2 Industrial pollution

The implementing decrees on this question have been particularly slow in appearing. The most important is the Royal Decree of 8 August 1975 relating to the prevention of atmospheric pollution by sulphur oxides and dust arising from industrial combustion installations.[19]

This Decree is applicable to combustion installations which are not subject to the provisions of the Royal Decree of 16 July 1971, or to those of the Royal Decree of 26 March 1971, but it does not apply to internal combustion engines or to explosion and combustion processes which form an integral part of chemical or metallurgical processes.

A distinction is made between traditional thermal electric power stations and other installations.

3.3.2.1 TRADITIONAL THERMAL ELECTRIC POWER STATIONS

With regard to thermal electric power stations which run partially or totally on *coal,* the gas emitted from the chimneys must not contain more than 500 mg of dust per Nm^3 [20] for power stations which run on coal containing 20% or more of ash, and 350 mg/Nm^3 for power stations which run on coal containing less than 20% ash. The use of anti-dust control devices is compulsory.

With regard to electric power stations which run totally or partially on *liquid fuel,* the maximum permissible content of sulphur dioxide in the smoke emissions must not exceed 5 g/Nm^3, from 1 January 1980. But each time the meteorological conditions appear unfavourable for the dispersion of the combustion gases—which in the first instance should be recognised by the operator 'by the use of local measuring apparatus'—the concentration of sulphur dioxide in the emission gases must be reduced to a value of less than or equal to 2 g/Nm^3. To this end, the user must keep available a sufficient stock of fuel with a low sulphur content to permit the plant to function for seven consecutive days. Finally, the Bacharach smoke index for the emissions must not exceed 4 outside lighting up or fuel changing periods.

With regard to the dispersion of gaseous pollutants, the Royal Decree of 8 August 1975 imposes a minimum chimney height to be calculated according to a formula contained in article 9. In the specially protected zones, the chimneys must be of a height equal to those outside the special zones multiplied by a coefficient of 1.4. The authority responsible for issuing orders under the regulation on classified installations (see Chapter

2) can always impose a chimney height greater than that envisaged by the formula in the Royal Decree, if it is a question of a new electric power station of more than 100 MW.

In addition to these measures relating to emissions, there are obligations relating to immission levels. When the total capacity of an installation is more than 200 MW on the same site, the operator must install apparatus to measure the ground level concentrations of sulphur dioxide. Each time the average air concentration in 24 hours of sulphur dioxide reaches a value exceeding 500 μg/m^3, the operator must take the necessary steps to reduce the sulphur dioxide concentration in the emission gas to less than or equal to 2 g/Nm^3.

3.3.2.2 OTHER INDUSTRIAL COMBUSTION INSTALLATIONS

The controls vary according to whether the installation is located in a specially protected zone or not.

The gaseous emissions from the chimneys of installations which run on combustible fuels must not contain more than:

150 mg of dust per Nm^3 (specially protected zones);

300 mg of dust per Nm^3 (outside the specially protected zones).

The combustion of fuels in a series of apparatuses linked to the same chimney must be constructed in such a way as to ensure that the sulphur dioxide emission does not exceed:

(a) for a calorific capacity not exceeding 1,000,000 kcal/h, from 1 October 1980: 0.85 g/Nm^3;

(b) for a calorific capacity of more than 1,000,000 kcal/h, outside the specially protected zones: 4.7 g/Nm^3; in the specially protected zones: 3.2 g/Nm^3, if the calorific capacity is less than or equal to 20,000,000 kcal/h, and 3.7 g/Nm^3 if the calorific capacity is greater than 20,000,000 kcal/h.

In the specially protected zones, the combustion of peat, lignite and non-smokeless agglomerates is forbidden.

Each time the meteorological conditions appear unfavourable for the dispersion of the combustion gas, bearing in mind the location, the operator of a series of combustion units of more than 100 kcal/h, running wholly or partially on liquid fuels, must take the necessary steps to reduce the sulphur dioxide content of the emission to a value of less than or equivalent to 2 g/Nm^3.

To ensure a suitable dispersion of the gaseous pollutants, the Royal Decree also imposes certain standards relating to chimney height.

Finally, there is a specific obligation for the chief operator of an oil refinery to install and maintain measuring equipment in the area surrounding the refinery to monitor and ensure that the average 24 hourly concentrations of sulphur dioxide in the air do not exceed 500 µg/m³.

3.3.2.2.1

The Royal Decree of 8 August 1975 also envisages a desulphurisation programme for liquid fuels. At present they cannot be placed on sale in Belgium—at least with regard to industrial combustion installations other than traditional electric power stations—unless their sulphur dioxide content is less than or equivalent to a specific percentage:[21]

gasoils	0.3%
light fuel oils	0.5%
intermediary fuel oils	1.3%
heavy fuel oils	1.9%
extra-heavy fuel oils	2.2% (specially protected zones) 2.8% (outside specially protected zones)

3.3.2.2.2

Purification measures which require a considerable financial contribution can sometimes be granted a derogation by ministerial decree, for a maximum duration of three years which can be renewed. In fact, the only exception has been the Ministerial Decree of 13 September 1976 relating to greenhouses in the horticultural sector and to mushroom growing farms, which envisaged the gradual application of purification measures over a period of three years.[22]

3.3.3 Pollution from vehicles

Within the framework of the powers granted to him by the Law of 28 December 1964, a number of decrees have been issued by the King to limit atmospheric pollution from motor vehicles:

(a) The Royal Decree of 8 November 1971[23] to limit the carbon monoxide content of exhaust gas emitted by motor vehicles equipped with controlled ignition. These vehicles must be controlled and maintained in such a manner as to ensure that, at low speed, the exhaust gases emitted do not contain more than 4.5% of carbon monoxide.

(b) The Royal Decree of 25 July 1975 relating to the type approval of motor vehicles with controlled ignition with regard to the emissions of gaseous pollutants by the engine[24] establishes the standards for exhaust gases which must be satisfied in order for the motor vehicle to be approved.

These standards are laid down in regulation no. 15 attached to the agreement concerning the adoption of uniform type approval conditions and the reciprocal recognition of the type approval of equipment and parts of motor vehicles, signed in Geneva on 20 March 1958, and the Directive of the Council of the European Communities of 20 March 1970 concerning the approximation of the legislation in the Member States relating to the measures to be taken against air pollution by gases arising from engines with controlled ignition in motor vehicles,[25] amended by the Directive of the Council of the European Communities of 28 May 1974, of 30 November 1976 and of 14 July 1978.[26]

(c) The Royal Decree of 11 January 1977 relating to the type approval of diesel engined vehicles[27] fixes the conditions for gas exhaust which must be satisfied by these vehicles. The standards applied are those of regulation no. 24 attached to the agreement signed in Geneva on 20 March 1958, concerning uniform requirements relating to the type approval of vehicles with diesel engines as regards the emission of pollutants by the engine, modified by the amendment of 11 September 1973, and by those contained in the Directive of the Council of the European Communities of 2 August 1972 concerning the unification of the legislation of the Member States relating to measures to be taken against the emission of pollutants arising from diesel powered engines intended for the propulsion of vehicles.[28]

(d) The Royal Decree of 26 February 1981[29] implements the EEC directives concerning a harmonised type approval procedure for motor vehicles and their trailers, wheeled agricultural or forestry tractors, their components and their security devices.

(e) The Royal Decree of 29 March 1977, as amended by a Royal Decree of 21 April 1982, limits the content of lead in petrol.[30] At present it is forbidden to import for consumption, to distribute or to sell petrol with a lead content, calculated as lead, exceeding 0.4 g/l.

3.4 ORDINARY LAW

The promulgation of the Law of 28 December 1964 has not had the effect of erasing the past. Specific texts which are of certain relevance should therefore be mentioned.

3.4.1 The Penal Code

3.4.1.1

The Penal Code contains a whole series of regulations which can be applied in certain cases of atmospheric pollution; however, the inadequacy of the punishments reduces the effectiveness and intimidating character of these legal ancestors of the Law of 28 December 1964 to very modest proportions.

(a) *Article 552(1)* punishes—with a fine of from one franc to ten francs (which is ridiculously little, even when multiplied by six)—'whoever throws, displays or abandons on a public highway objects of a nature to cause harm by their fall or *by noxious exhalations*'.

Even though this text represses abandonment, voluntary or involuntary, direct or indirect, committed by man,[31] it only covers objects which are displayed or abandoned on the public highway and 'cannot apply when the objects which cause harm because of their noxious exhalations are displayed or abandoned on private property even though their noxious exhalations reach the public highway'.[32]

(b) *Article 552(5)* punishes—with the same futile fine—those who, imprudently, throw on to an individual 'any kind of object which can inconvenience or soil'.

(c) *Article 557(4)* punishes—with a fine of from five francs to fifteen francs and imprisonment of from one day to four days, or only one of these punishments—those who have thrown objects which could soil or degrade, against cars or buildings.[33]

(d) *Article 559(3)* punishes—with a fine of from ten francs to twenty francs—those who, through carelessness or lack of precaution, kill or injure an animal, by throwing substances of any kind.

(e) *Article 563(3)* punishes—with a fine of from fifteen francs to twenty-five francs and imprisonment of from one day to seven days, or one of these punishments only—those who voluntarily, but without intention to injure, throw any object at someone which causes inconvenience or soiling.

3.4.1.2

Very judiciously, H. Bocken points out that these regulations are not of interest from the point of view of penal law; they are only of real importance in that, on the civil law level, they allow a certain number of atmospheric pollution cases to qualify as involving fault, without the need to prove carelessness or lack of precaution. In this way, they can, if the occasion arises, enable injured persons to constitute themselves as a 'civil party' in a penal action and thus start a civil action which could lead to the implementation of measures to put an end to the atmospheric pollution.[34]

3.4.2 General Regulation for the Protection of Labour

As has already been indicated, the control of classified installations (Title I of the General Regulation for the Protection of Labour) is—and remains—applicable to all installations included in the list attached to Title I.

This General Regulation for the Protection of Labour also contains a whole range of particular requirements which are applicable to certain industries (Title III). It refers in particular to such diverse industries as the treatment of minerals, zinc and lead, brick manufacturing, bakeries and confectioners.[35,36]

3.4.3 General Regulation on Road Traffic Policing

With regard to pollution from motor vehicles, it should be noted, in addition to the executive decrees of the Law of 28 December 1964 already mentioned (see 3.3.3) that some requirements under the Royal Decree of 1 December 1975 act as a general regulation on road traffic policing.[37]

As a result of article 7, it is prohibited to hinder or endanger the circulation of traffic, particularly by spreading smoke or steam over the public highway; this is a punishable offence, even if caused by negligence.[38]

More important is the requirement under article 81(3)(1), which prohibits in particular:

(1) With the exception of transient emissions of smoke caused during the starting up of an engine or the gear changing of a vehicle, the production of smoke which exceeds the limits fixed by the technical regulation on motor vehicles.[39]

(2) The emission of polluting gases which exceed these same limits.

3.4.4 Other legislation

It goes without saying that other legislation can have an influence on the problems created by atmospheric pollution.

This is particularly true of the land use and town planning legislation of 29 March 1962, on the basis of which building permits can impose certain conditions on the construction of dwellings or industrial installations, and the legislation on economic expansion, on the basis of which the granting of subsidies or other forms of State aid can be subject to the observance of certain conditions which are aimed at the prevention of atmospheric pollution.[40]

3.4.5 Communal regulations

In a decision of 10 July 1973, the Council of State solved the thorny and hotly debated question of whether or not the Law of 28 December 1964 excluded communal regulatory responsibility on the subject.

The Council of State remarked that the fact that there is currently a law empowering the King to take all appropriate measures does not in itself justify the conclusion that a subject that the Communal Council is competent to control as a result of its powers under article 50 of the Decree of 14 December 1789, article (3), Title XI of the Decree of 16–24 August 1790 and articles 75 and 78 of the Communal Law, would cease to be of Communal interest.[41]

Notes

1. J. Matthijs, La protection pénale de l'environnement. L'eau, l'air, le bruit, *Rev. dr. pen. crim.*, 1972, p. 544.
2. See, however, Chapter 1.
3. Baron Jean Constant, La protection pénale de l'environnement en droit Belge, in *Rapports belges au Xᵉ Congres international de droit comparé* (Belgian reports to the Tenth Congress on International Comparative Law), 1978, p. 602.

4. *Moniteur belge,* 23 October 1971. The working rules of the Commission were approved by Ministerial Decree of 9 July 1974, *Moniteur belge,* 13 November 1974.
5. *Moniteur belge,* 3 October 1972, 22 May 1976, 22 October 1976, 22 June 1978.
6. Royal Decree of 13 December 1966, modified by Royal Decree of 27 May 1968, fixes the conditions and methods of approval for laboratories and organisations in charge of sampling, analyses, tests and research within the framework of the fight against atmospheric pollution (*Moniteur belge,* 14 February 1967 and 13 July 1968). Approval can be limited to one or several emissions envisaged by the law or to a specified object within the framework of these emissions. It can be totally or partially, temporarily or permanently withdrawn if the conditions imposed are not observed.
7. As a result of article 36 of the Law of 25 July 1981, the amount specified for the fines should be multiplied by 40, in order to bring it into line with the depreciated value of money.
8. Royal Decree of 26 July 1971 relating to the creation of specially protected zones against atmospheric pollution (*Moniteur belge,* 5 August 1971), amended by Royal Decree of 3 July 1972 (*Moniteur belge,* 26 September 1972), and by Royal Decree of 29 January 1974 (*Moniteur belge,* 2 March 1974).
9. Article 6, Royal Decree of 26 July 1971, amended by article 1 of 3 July 1972. The requirements of this article do not prejudice the application of other regulations and particular decrees in force, especially the Communal Regulations.
10. *Moniteur belge,* 10 June 1978.
11. *Moniteur belge,* 8 May 1971.
12. *Moniteur belge,* 9 March 1978.
13. An installation is considered to be in good working condition if it satisfies the following conditions:

    (1) Installations running on solid fuel:
        (a) not to emit visible smoke except on rare and transient occasions;
        (b) to have an average weight index of combustion gas less than or equivalent to 0.6 g/1000 kcal produced in the furnace.

    (2) Installations running on liquid fuel:
        (a) not to emit smoke particles at any time;
        (b) to have a black smoke index less than or equivalent to 3;
        (c) to be controlled in such a way that the filter paper used during trials to measure the black smoke index gives no visible trace of oil;
        (d) to be controlled in such a way that the gases at the generator exit have a CO_2 content exceeding 9% in dry gas volume and a temperature less than 300°C above the ambient temperature in the stoke hole (article 3).
14. For the details of this maintenance, see article 4 at the end of the Royal Decree dated 6 January 1978.
15. The methods for carrying out control tests are established in great detail by Annex I of the Royal Decree of 6 January 1978.
16. The recognition application is submitted to the Nuisance Service of the Public Hygiene Administration, Ministry of Public Health and the Family.
17. The aptitude certificate and permanent qualification certificate are awarded to individuals who have completed a training course and passed an examination described in article 20, in an establishment approved for the purpose by the Minister of Public Health. This approval is only given to establishments which satisfy the conditions listed in the Royal Decree (article 20 and following). Approval can be withdrawn by decree.
18. 'Appropriate organisation' is taken to mean an organisation approved by the Minister of Public Health to carry out admission verifications for heating installations in accordance with the requirements of the Royal Decree of 13 December 1966. See note 6.

19. *Moniteur belge*, 2 October 1975.
20. Nm^3 = cubic meter of gas at 0° and 760 mm mercury, the water remaining in the form of vapour.
21. See the Directive of the Council of the European Communities of 24 November 1975, OJ L307, 27 November 1975.
22. *Moniteur belge*, 21 December 1976.
23. *Moniteur belge*, 11 November 1971.
24. *Moniteur belge*, 27 August 1975. See also no. 26.
25. *Official Journal of the European Communities*, L76, 6 April 1970.
26. *Official Journal of the European Communities*, L159, 15 June 1974, OJ L32, 3 February 1977 and OJ L223, 14 August 1978.
27. *Moniteur belge*, 22 February 1977.
28. *Official Journal of the European Communities*, L190, 20 August 1972.
29. *Moniteur belge*, 1 May 1981.
30. *Moniteur belge*, 7 July 1977 and 27 May 1982. The Royal Decree of 21 April 1982 implements Directive 78/611/EEC of 29 June 1978 of the Council of the European Communities on the approximation of the laws of Member States concerning the lead content of petrol. The lead content of petrol is measured using the method described in Belgian Standard NBN T52.111 (First edition, June 1978).
31. Court of Cassation, 5 March 1951, *Pas.*, 1951, I, 437–439.
32. J. Constant, *op. cit.* p. 590; and Manuel de droit pénal, deuxième partie, *Les Contraventions*, Volume II, p. 590, no. 1754; Rigaux and Trousse, Code de Police, Volume I, p. 63.
33. However, the commercial tribunal of Antwerp, in a judgement on 27 June 1899, refused to apply this requirement to the smoke emissions caused by a factory; *J. T.* 1899, p. 859.
34. See H. Bocken, Juridische aspecten van de bestrijding van de luchtverontreiniging, *T.B.P.*, 1974, p. 21.
35. For a critical and detailed analysis of the general regulation for protection at work, see Monique Sojcher-Rouselle, *Droit de la securité et de la santé de l'homme au travail*, Brussels, 1979.
36. See also article 36 of the Royal Decree of 28 February 1963, acting as a general regulation for the protection of the population and of the workers against the danger of ionising radiations: 'The emission of radioactive substances into the atmosphere in the form of gas, dust, smoke or vapour is forbidden if the concentration of radioactive substances at the point of emission into the atmosphere is greater than one-tenth of the permissible maximum envisaged for inhaled air in the annex to this Chapter'. See Chapter 8.
37. *Moniteur belge*, 9 December 1975.
38. Court of Cassation, 28 November 1960, *Pas.*, 1961, I, 332.
39. See article 39 of the Royal Decree of 15 March 1968 acting as a general regulation on the technical conditions to which motor vehicles and their trailers must conform, amended by Royal Decree on 14 January 1971, 9 August 1971 and 12 December 1975, *Moniteur belge*, 28 March 1968, 20 January 1971, 2 September 1971 and 30 December 1975. See also the Royal Decree of 1 July 1964 fixing the conditions with which diesel powered motor vehicles must comply with regard to smoke emissions, *Moniteur belge*, 17 July 1964.
40. H. Bocken, *op. cit.* p. 26.
41. See Council of State, 10 July 1973 no. 15.973; Commune of Etterbeek, *Mouvement communal*, 1974, p. 194 with note by J. Blanquart.

4
Noise

4.1 THE LAW OF 18 JULY 1973 RELATING TO THE CONTROL OF NOISE

4.1.1 Introduction

The Law of 18 July 1973 is characterised both by its very general nature and its vast field of action: it applies particularly to noise caused by 'motor vehicles—lorries, cars, motorbicycles, motorcycles, aeroplanes, helicopters, railways, warning signals at unsupervised level crossings, boats, machinery installed in factory workshops, machines installed in building yards and household equipment (article 1, at the end).

4.1.2 Powers

In the interest of the health of individuals, the King is empowered to take any necessary measures to prevent or control noise arising from mobile or stationary, permanent or temporary sources. Specifically, he can:

(1) prohibit the production of certain noises;
(2) subject the production of certain noises to restrictions, including limiting the time during which the noise can be produced;
(3) control or prohibit the import, manufacture, export, transit, transport, offer for sale, sale, conditional or free offer, distribution, installation and use of equipment, devices or objects producing or likely to produce certain noises;
(4) impose and control the siting and use of equipment or devices

intended to reduce, absorb or put right the inconvenience arising from noise;

(5) create protected zones for which specific measures can be taken.

The King can equally impose technical building or installation conditions to reduce the spread of noise and its inconvenience.

In particular the King can impose technical conditions on the construction of new roads, railways, airfields, extensions to existing roads, railways or airfields and on the implementation of regional development plans or specific plans.

Finally, the King can impose special conditions concerning professional qualifications and access to the profession for personnel who could be in charge of the installation or maintenance of devices for the purpose of reducing the production of noise.

4.1.3 Execution of the law

The royal decrees in execution of the law must be submitted to the Superior Council of Public Hygiene. They are jointly proposed by the Minister of Public Health and the Environment and the various ministers concerned (seven in all).[1]

The Minister for Public Health and the Environment alone is able to propose decrees concerning sources of noise which, because of their origin, do not come under one of the other ministerial departments mentioned in the law.

This same Minister is in addition responsible for co-ordinating and promoting scientific research and studying methods of effectively combating noise, in collaboration with individuals, laboratories and organisations approved for this purpose by him.[2]

Finally, the Minister for Public Health and the Environment can promote the education of the population on the subject of the control of noise; he can make suggestions to the Ministers for Education for the introduction of such subjects into the education programmes.

4.1.4 Monitoring the application of the law

The King will designate specially skilled agents to monitor the application of the law and its executive decrees, but, as is the case relating to atmospheric pollution, to date there is no royal decree designating agents with general responsibility on the subject. This also applies to judiciary police officers—these agents search out and verify any contraventions committed. In the absence of proof to the contrary, the written reports they prepare are taken as verification of the facts stated therein. This applies not only to the information collected through the use of measuring equipment but equally to all data obtained by any other legal means. While it is true that this regulation offers little legal guarantee to subjects under its jurisdiction, it has been considered necessary to allow flagrant contraventions to be pursued, even at a time when the agent has no sonometer to hand.

These agents can enter all establishments, day and night—with the exception of residential dwellings—if they have reason to believe that the law and its implementing decrees are being contravened.

If sufficient indicators exist to presume that the origin of an atmospheric pollution incident is to be found in a residential building, a house visit can be made between 5.00 am. and 9.00 pm. by two of these agents, acting on the authorisation of a police tribunal judge.

In addition, these specially designated agents can test, or cause to be tested by approved individuals, laboratories, public or private organisations,[3] the equipment or devices liable to produce noise or intended to reduce, absorb or put right the inconvenience of noise; these tests must always take place in the presence of the interested party or a representative appearing on his behalf.

Finally, the specially designated agents can provisionally prohibit the use of equipment and devices which, because of their construction or characteristics, are not in an operating condition which conforms with the executive decrees of the law; alternatively, they can seal off such equipment or devices and take any emergency measures which the situation demands in the interest of the public and cleanliness.

These measures cease to have effect after a period of eight days if, during that time, they have not been ratified by the official in charge of the administration to which the agent belongs, who must previously hear or at least call the users.

The users of the equipment and devices are notified at once by registered letter of all ratification decisions. An appeal to the King against ratification decisions is open to all interested parties. The King controls the methods of appeal[4] which has no suspensive effect.

The specially designated agents can request the assistance of the communal authorities in the carrying out of their duties.

4.1.5 Sanctions

Article 11 of the law punishes the following with imprisonment of from eight days to six months and with a fine of from 26 francs to 5,000 francs,[5] or with one of these punishments only:

(1) The holder of equipment or devices which, as a result of negligence or lack of foresight on their part, give rise to the form of noise prohibited by the King.
(2) Those who infringe the requirements of the royal decrees executing the present law.
(3) Those who refuse or oppose inspection visits, tests or measures described in 4.1.4.

The punishments can be doubled if a new contravention is committed within two years from the date of a previous conviction for one of the contraventions contained in this article and having the force of a judgement.

4.1.6

The Law of 18 July 1973 expressly states in articles 12 and 13 that it does not prejudice the provisions of the Law of 10 June 1952 concerning the health and safety of workers and hygiene at work, the system of control of classified installations, the General Regulation for the Protection of the Population and Workers against the Danger of Ionising Radiation, or the communal regulations issued as a result of the decrees of 14 December 1789 and 16–24 August 1790.

4.2 IMPLEMENTING DECREES OF THE LAW OF 18 JULY 1973

The Law of 18 July 1973 was followed by executive decrees in three sectors:

(1) organisation of races, training and trials for motor vehicles and motorcycles;
(2) music in public and private establishments;
(3) subsidies for the purchase of sonometers by the decentralised authorities.

4.2.1 The Royal Decree of 10 June 1976 controlling the organisation of races, training and trials for motor vehicles[6]

This decree is based on two principles:

(1) the straightforward prohibition of races, training and trials in certain zones;
(2) the obligation to obtain authorisation for the organisation of these activities outside the protected zone.

A distinction is made between the circuits and sites used in a permanent fashion and those which are not for permanent use.

Circuits or areas used in a permanent fashion are taken to mean: 'circuits or sites for recreational use and/or on which training and trials are organised and/or on which more than one race a year, as well as related training, are held' (article 2(2)).

If, instead, there is a maximum of only one race a year, and related training, then the circuit or site is considered to be used in a non-permanent fashion.

4.2.1.1 GENERAL PROHIBITION: PROTECTED ZONES

Speed races, obstacle races, trials and training are prohibited on the following circuits or sites

(a)	(b)
used in a *non-permanent* fashion located 350 m from the limit of:	used in a *permanent* fashion located 500 m from the limit of:

(1) an agglomeration as defined in the general regulation on the policing of road traffic;
(2) establishments requiring silence: nursing homes, rest homes, scientific and educational establishments, cultural centres, public libraries and museums;
(3) regions classified as nature reserves, national parks or forestry reserves, as defined in the Law of 12 July 1973 on nature conservation;
(4) natural areas of scientific interest or nature reserves, as defined in the Royal Decree of 28 December 1972 relating to the presentation and implementation of draft plans and regional development plans;
(5) beaches, camp sites, residential camping parks and residential week-end parks;

(6)   public parks.

This distance is reduced to 200 m for circuits and sites mentioned under letter (a) and to 300 metres for those under letter (b) when the motor vehicles or motorcycles used conform to the requirements fixed by the highway code with regard to the question of sound level.[7]

4.2.1.2 AUTHORISATION: CIRCUITS OR SITES FOR PERMANENT USE

Races, trials and training are only permitted with authorisation from the Minister of Public Health and the Environment.

The authorisation application must be presented by the site operator to the Minister at least three months before the first proposed activity. It must contain the following information:

(1)   the identity of the applicant;
(2)   the identity of the owner of the circuit or site; if the applicant and the owner are two different people, the application must be accompanied by a written agreement from the owner;
(3)   a map of the circuit or site (scale 1/2,500), indicating devices for noise reduction (natural or artificial);
(4)   a map of the neighbourhood surrounding the circuit or site (radius of 2 km, scale 1/25,000) indicating the elements envisaged in article 3 of the present decree;
(5)   the nature of the activities and frequency of use of the circuit or site;
(6)   characteristics of vehicles used (type, engine capacity, noise level);
(7)   calendar of activities during the first three months.

Before granting an authorisation, the Minister must ask for an opinion from a 'Consultation Commission'.[8]

The Commission gives its opinion within two months from receiving the request. The opinion may contain approval conditions, especially concerning the equipping of the site or circuit and the establishment of buffer zones and restrictions relating to the use of the circuit or site.

The authorisation mentions the private or public person or organisation to which it is granted, as well as the location and geographic situation of the site or circuit, the conditions and restrictions to which the use of the site or circuit is subject and the calendar of planned activities for a period of at least three months.

It should be emphasised here that for circuits or sites already in existence before 1 June 1974 and equipped and managed to this effect, the Minister can derogate the general prohibition and authorise races, trials and training even when these circuits or sites are located less than 500 m— or 300 m—from the border of zones requiring quiet as listed above under 4.2.1.1.

4.2.1.3 AUTHORISATION: CIRCUITS OR SITES FOR NON-PERMANENT USE

Races and related training are only permitted with authorisation from the deputed states of the province in which the circuit or site is located.

The authorisation is only granted in consultation with the provincial hygiene inspector or the inspector of nuisances.

The deputed states can, in principle, derogate the prohibition and authorise races on circuits located less than 350m—or 200 m—from the border of zones requiring quiet, in the case of circuits or sites which were in use before 1 June 1974. In this case, approval from the Consultation Commission is required.

The organiser formulates an application for authorisation in writing and sends it, at least three months in advance of the date for the planned activities, to the Governor of the Province in which these activities would take place.

The application must include:

(1) the identity of the applicant;
(2) the date, time, duration and nature of the activities;
(3) the characteristics of the vehicles used (type, engine capacity, sound level);
(4) a map of the site or circuit and of the neighbouring area (radius 500 m, scale 1/10,000) indicating the elements envisaged in article 3 of the present decree.

If the applicant wishes to appeal for the possibility of derogation, the application must be presented at least four months before the date of the planned activities and, in addition, must contain proof that the circuit or site was already in use several years before 1 June 1974, and also justify the request for derogation, with a description of devices (natural or artificial) for the limitation of noise and an indication of these devices on the map.

Within two months from receipt of the complete application file, the Governor informs the applicant of the deputed states' decision on authorisation.

This period is extended to three months for applications involving derogation.

The authorisation mentions the person or private or public organisation to which it is granted as well the date, time, duration and nature of the activities, the characteristics of the vehicles used and the location and geographic situation of the site or circuit.

If the activity mentioned in the authorisation cannot take place on the date specified, the deputed states can alter the date contained in the authorisation without asking the opinion of the provincial hygiene inspector, the inspector of nuisances or the Consultation Commission.

4.2.1.4 CONTROL AND SANCTIONS

The specially designated agents to monitor the application of the requirements of this royal decree and to ascertain any contraventions are officials of the Hygiene Inspection and Nuisance Services of the Ministry of Public Health.

When races, trials and training controlled by this royal decree take place with motor vehicles and motorcycles which conform to the standards fixed in the highway code relating to noise level—in such a way that they can take place at a shorter distance from zones requiring quiet— the organisers of these activities are obliged to determine, or cause to be determined, the noise level of each of the vehicles and to record the results of these sonic measurements in a register which must be kept available for inspection by the control officials in charge.

4.2.2 The Royal Decree of 24 February 1977 fixing acoustic standards for music in public and private establishments[9]

4.2.2.1

This royal decree contains two series of provisions:

(1) those which control the maximum sound level inside public establishments;
(2) those which control the sound level in the neighbourhood of public and private establishments in which music is produced.

4.2.2.2

Music is taken to mean all the methods of emission of electronically amplified music arising from permanent or temporary sources (article 1). It is therefore a question of music performed in front of a microphone and emitted through loudspeakers.[10]

While it is clear that the royal decree does not apply to music produced by fanfares, orchestras or the ringing of bells,[11] it is questionable whether or not this is the case for music produced by electric guitars or electric organs if it is not emitted through a microphone.

4.2.2.3

In *public establishments* the maximum sound level emitted by music must not exceed 90 dB(A). This sould level can be measured in any part of the establishment where people would normally be found.[12]

All establishments and their annexes which are accessible to the public, even if access is restricted to certain categories of people, whether by payment or not, as well as dance halls, concert halls, discotheques, private clubs, shops, restaurants and wine bars, including those situated outside, are considered to be public establishments.

The sound level in dB(A) is measured by a sound meter which must satisfy the conditions of precision defined in the Belgian Standard NBN 576.80 (Belgian Institute of Standards) with a 'slow' dynamic characteristic.

Before each measurement or series of measurements relating to the same noise source is taken, the sound meter must be correctly calibrated with the help of a standard acoustic source.

4.2.2.4

Public and private establishments in which music is produced must be fitted out in such a way as to ensure that the sound level measured in the neighbourhood:

(1)   does not exceed 5 dB(A) above the background noise, when this is less than 30 dB(A);
(2)   does not exceed 35 dB(A) when the level of the background noise is between 30 and 35 dB(A);
(3)   does not exceed the level of the background noise when this is over 35 dB(A).

'Level of background noise' is taken to mean the minimum sound level measured during a period of five minutes, excluding 'music' in the meaning of article 1 of the royal decree.

This sound level is measured inside a room or building with the doors and windows closed. The microphone is placed at least one metre away from the walls at a height of 1.20 m above the floor.

The royal decree does not, therfore, protect the peace and quiet of the gardens and courtyards of the neighbourhood, but only the insides of rooms and buildings in the neighbourhood. It has been justifiably observed that neighbours wishing to sleep with their windows open are not protected either.[13]

4.2.2.5

The search for and verification of contraventions of the provisions of the Royal Decree of 24 February 1977 are the responsibility of the following officials—without prejudice to the powers of the judiciary police officers:

(1) technical officials of the Nuisances Service and hygiene inspectors of the Ministry of Public Health and the Family;
(2) technical officials of the Department of the Environment, and of the Institute of Hygiene and Epidemiology of the Ministry of Public Health and the Family;
(3) responsible officials from the Administration of Safety at Work of the Ministry of Employment and Work, with regard to dangerous, dirty or noxious estahlishments under their supervision;
(4) technical officials of provinces, agglomerations of communes and communes designated by a royal decree (which has not yet come into force).

4.2.2.6

During the months of May and June 1977 the consumer magazine *Test Achat* proceeded with acoustic meeasurements of concerts of pop, rock, jazz and other music, as well as in night clubs, in order to ascertain to what extent the imposed limit of 90 dB(A)[14] is respected. At all the concerts visited in this way, the imposed standaid was exceeded, although a number of night clubs did comply with the new regulation.

This investigation, therefore, clearly demonstrated that the majority of public establishments take no notice of the standards imposed.

4.2.3 The Royal Decree of 18 May 1977 fixing the conditions of approval and the percentages of grants available for the purchase of sound meters for the provinces, agglomerations of communes and communes[15]

The Minister of Public Health and the Environment can grant to provinces, agglomerations of communes and communes, subsidies for the purchase of sound meters and standard sources for the purpose of combating noise provided that these sound metres satisfy the following conditions:

(1) they must conform to the precision criteria defined in the Belgian Standard NBN 576.80;
(2) they must be equipped with an (A) circuit balancer;
(3) they must be equipped with 'slow and fast dynamic characteristic' devices;
(4) they must be suitable for measuring sound levels of between 30 and 110 dB (A).

The subsidy is granted, in principle, for the purchase of one sound meter per 50,000 inhabitants; the subsidy can be up to 80% of the expenditure, up to a maximum of 60,000 Belgian francs, including V.A.T.

4.3 ROAD TRAFFIC NOISE

4.3.1

Article 81.3.1 of the General Regulation on the Policing of Road Traffic states that motor vehicles must be conditioned, maintained and operated in such a manner as not to endanger traffic safety or inconvenience other road user. To this end, it is forbidden to inconvenience the public or to frighten animals by noise; in no case can the sound level exceed the limits fixed by the technical regulations on motor vehicles, motorbikes and autocycles.

Article 40 of the Royal Decree of 15 March 1968 acts as a General Regulation on the technical conditions to which motor vehicles and their trailers must conform with regard to the noise emitted by the motor vehicles in operation. This noise must not exceed the following limits:[17]

Category	Values expressed in dB (A)
1. Cars, mixed use cars and minibuses	84
2. Coaches and buses with a maximum authorised weight of less than 3,500 kg	86
3. Vehicles for the transport of goods with a maximum authorised weight of less than 3,500 kg	86
4. Coaches and buses with a maximum authorised weight of over 3,500 kg	91
5. Vehicles for the transport of goods with a total authorised weight of, over 3,500 kg	91
6. Coaches and buses with an engine capacity equal to or exceeding 200 CV DIN	93
7. Vehicles for the transport of goods with an engine capacity equal to or exceeding 200 CV DIN and with a maximum authorised weight of over 12,000 kg	93

With regard to motorbikes and autocycles the noise produced must not exceed the following levels:[18]

Category	Values expressed in dB(A)	
	Vehicles in use	New vehicles (since 1 January 1982)
Class A motorbikes	85	70
Class B motorbikes	90	75
Two-wheeler motorcycles with or without sidecars with a cylinder capacity of up to 50 cm³	95	80
Two-wheeler motorcycles with or without sidecars with a cylinder capacity of between 50 cm³ and 125 cm³	97	82
Two-wheeler motorcycles with or without sidecars with a cylinder capacity of between 125 cm³ and 500 cm³	99	84
Two-wheeler motorcycles with or without sidecars with a cylinder capacity over 500 cm³	101	86
Three-wheelers	99	84
Permitted tolerance	3	1

4.3.2

Article 33 of the General Regulation on the Policing of Road Traffic prohibits the use of acoustic warning devices other than those envisaged by the Regulation or by the technical regulations on motor vehicles, motorbikes and autocycles. Acoustic warnings should be kept as brief as possible. They are only permitted to give necessary warning in order to prevent an accident, or, outside built up areas, when it is necessary to alert another driver before overtaking. Between dusk and dawn, except in cases of imminent danger, acoustic warnings should be replaced by the brief and alternating use of main beam and dipped headlights.

4.3.3

Article 45(5) of the General Regulation on the Policing of Road Traffic makes the driver responsible for taking all necessary measures to ensure that loading, as well as all accessories used for packing or protection of loading operations, do not, as a result of the noise they produce, annoy drivers, inconvenience the public or frighten animals.

4.3.4

Two Royal Decrees of 3 December 1976 control the type approval of motor vehicles with regard to permissible sound level, of acoustic warning devices and of motor vehicles in relation to acoustic signalling.[19]

These regulations cover the conditions and methods under the Directive of the Council of the European Communities of 6 February 1970,[20] and by Regulation no. 28, attached to the Agreement concerning the adoption of uniform type approval conditions and reciprocal recognition of the type approval of equipment and parts of motor vehicles, signed in Geneva on 20 March 1958.

No motor vehicle can be the subject of type approval if it does not conform to the prescriptions of points 1 and 2 of the annex of the Directive of the Council dated 6 February 1970, amended by the Commission Directive of 7 November 1973.

4.3.5

The Royal Decree of 26 February 1981 implements the EEC Directives concerning a harmonised type approval procedure for motor vehicles and their trailers, wheeled agricultural and forestry tractors, their components and their security devices,[21] e.g. Council Directive 77/212/EEC of 8 March 1977 amending Directive 70/157/EEC on the approximation of the laws of Member States relating to the permissible sound level and the exhaust system of motor vehicles, and Council Directive 77/311/EEC of 29 March 1977 relating to the driver-perceived noise level of wheeled agricultural or forestry tractors.

4.4 AIR TRAFFIC NOISE

4.4.1

The Law of 27 June 1937 'revising the Law of 16 November 1919 relating to the control of aerial navigation', amended by the Laws of 4 August 1967 and 6 August 1973,[22] enables the King to decree 'all regulatory prescriptions concerning aerial navigation' (article 5).[3]

4.4.2

On the basis of this legislation the Ministerial Decree of 2 May 1975 fixed the noise standards for *subsonic aeroplanes*.

Since 1 January 1980 no Belgian navigability certificate is granted or renewed for a subsonic aeroplane—with the exception of aeroplanes which, at their maximum permitted weight, require a runway equal to or less than 450 m—unless the maximum noise level does not exceed the following values:[24]

(1) At the lateral measurement point and at the approach measurement point, 108 EPNdB for aircraft with a maximum authorised take off weight of equal to or more than 272,000 kg, decreasing by 2 EPNdB for each reduction in maximum take off weight by half, down to a value of 102 EPNdB for a maximum permitted weight equal to or less than 34,000 kg.

(2) At the measurement point flown over at take off, 108 EPNdB for aircraft with a maximum permitted take off weight of 272,000 kg or more, decreasing by 5 EPNdB for each reduction in maximum take off weight by half, down to a value of 93 EPNdB for a maximum permitted weight equal to or less than 34,000 kg.

If the maximum noise level is exceeded at one or two measurement points:

(a) the total excess must not be more than 4 EPNdB, or 5 EPNdB if it is an aeroplane with four jet engines, the dilution rate of which is

equal to more than 2 and for which the application for first certification in the country of manufacture was before December 1969;
(b) the excess at a given point must not be more than 3 EPNdB;
(c) compensation for any excess must be made by a corresponding decrease at one or other of the measurement points.

4.4.3

The Ministerial Decree of 2 May 1975 has been replaced by a Royal Decree of 5 June 1980,[25] implementing the Directive of the Council of the European Communities of 20 December 1979 on the limitation of noise emissions from subsonic aircraft.

As from 30 June 1986, no Belgian navigability certificate shall be granted or renewed for a subsonic aeroplane unless it complies with the requirements of the Royal Decree of 5 June 1980.

4.4.4

Article 4 of the Law of 27 June 1937 enables the King to prohibit the overflying of all or part of the territory by national or foreign aircraft.[24]

Articles 8, 56 and 63 of the Royal Decree of 14 May 1973 fixing the air regulations[27] establishes the minimum heights to be observed by aircraft and helicopters.[28]

In addition, the Administration of Air Routes vigorously applies the following provisions:[29]

(a) The prohibition of night flights: jet planes with a noise level exceeding the limits contained in 4.4.2 cannot land or take off at Belgian airports between 10.00 pm. and 5.00 am. G.M.T.
(b) The indication of runways to be used by preference for taking off and landing.
(c) The determination of taking off and landing conditions.

4.4.5

The Royal Decree of 15 March 1954 controlling aerial navigation,[30] subsequently amended several times, states, in article 21, that the flying ability of an aircraft inscribed in the aeronautical register must be verified by a navigability certificate granted by the minister in charge of aeronautical administration or by his deputy.

4.4.6

No civil aerodrome can be established without the authorisation of the minister in charge of aeronautical administration or his deputy. In addition, the opinion of the minister in charge of land planning and town planning, or his deputy, is required when the proposed aerodrome is of a permanent character.

The minister in charge of the aeronautical administration or his deputy fixes in each case the technical conditions for the use of aerodromes.

The captain of an aeroplane can only land on regularly established aerodromes except in the case of force majeure.

No modification can be made to an aerodrome unless the minister in charge of aeronautical administration has previously been informed. On this occasion, the minister can modify the conditions of use of the aerodrome.

4.4.7

Belgium has ratified the Convention Relating to Damage Caused to Third Parties on the Ground by Foreign Aircraft, signed in Rome on 7 October 1952, by the Law of 14 July 1966.[31] This law makes provision for the requirements of the Convention to apply to the Belgian territory, regardless of whether the aircraft is registered abroad or in Belgium.[32]

4.5 MACHINE NOISE

4.5.1

Article 148 *decies*, 2, of the General Regulation for the Protection of Labour contains the principle according to which 'all possible measures' must be taken to reduce sources of excessive noise and vibration arising from work or the workplace.

If the technical means for obtaining this reduction are found to be insufficient or inoperative, workers must use appropriate means for individual protection—ear plugs, ear muffs or protective helmets (article 157)—placed at their disposal by the employer. In that case, the employer must reduce the length of exposure time to these risks or introduce rest breaks during the working day.

4.5.2

On the basis of the Law of 11 July 1961 relating to safety guarantees which are obligatory for machines, parts of machines, materials, tools, apparatus and receptacles,[33] the Royal Decree of 26 September 1966 controls the construction, supply, maintenance and approval of sealing guns.[34]

4.5.3

Hyper acousis and deafness caused by noise are considered to be occupational illnesses which give rise to compensation, in accordance with the provisions of the Law of 24 December 1963 relating to compensation for damage resulting from occupational illness and its prevention.[35]

4.6 NOISE AND TOWN PLANNING

As already mentioned, as a result of article 2 of the Law of 18 July 1973, the King can impose technical conditions to reduce the inconvenience resulting from noise during the implementation of regional development plans or municipal plans. It appears from the debates in Parliament that a General Regulation was envisaged which would stipulate that development plans must make provision for areas planted with trees for the purpose of reducing noise intensity.[36]

These 'technical conditions' would tend to prevent both the penetration and emission of sound by acoustic insulation of dwellings. The legislation section of the Council of State has justifiably observed that these conditions should preferably be included in the general building regulations contained in article 59 of the Law of 29 March 1962.[35]

Be that as it may, in practice, there is no general regulation on the subject.

4.7 COMMON LAW

4.7.1 Disturbance at night

Article 561 of the Penal Code punishes those guilty of making a noise or disturbance at night which may disrupt the peace and quiet of residents, with a fine of from ten to twenty francs—to be multiplied by sixty—and by imprisonment of from one day to five days, or by only one of these punishments.

The contravention covered by article 561 of the Penal Code requires a personal act to be committed by the accused, be it voluntary, i.e. intentional, or involuntary, i.e. simply in error.[38]

Noise and disturbance at night which are purely the result of the normal carrying out of an occupation are not punishable.[39] However, the decision of the judge cannot be legally justified if, in order to acquit the operator of a commercial establishment of a contravention under article 561(1) of the Penal Code, he confines himself to ascertaining that the operation as it is organised and run has only resulted in noise at night which is of necessity linked to this operation and takes no account of obligations towards the neighbourhood or of the rights of neighbours to enjoy their rest at night.[40]

4.7.2 Communal regulations

4.7.2.1

As a result of the old Decrees of 14 December 1789 and 16–24 August 1790, the communes are obliged to ensure peace and quiet by means of police regulations.

> It is in this way that the communes can therefore include in a police regulation activities generating noise of a nature to disturb the peace and quiet of the residents, such as the use of lawn mowers and other garden implements, televisions, radios, model aeroplanes, etc.[41]

It has been judged that the measures prescribed by the communal authorities can be extended to private property and are therefore not restricted in their application to public places.[42]

However, both as regards electronically amplified music and motor racing, training trials, the communes must henceforth confine themselves to the strict application of the standards set by the central authority.

4.7.2.2

The Association of Belgian Towns and Municipalities has elaborated a standard police regulation concerning the control of noise (see Annex I to this chapter).

ANNEX I

Standard Municipal Police Regulation Concerning Control of Noise

The Municipal Council,

In view of the Decree of 14 December 1789 concerning the constitution of the municipalities, especially article 50;

In view of the Decree of 16–24 August 1790 concerning the administration of justice, especially article 3 of title XI;

In view of the Municipal Law, especially articles 75 and 78;

Considering that it is the duty of the municipal authorities to safeguard public tranquility;

Considering that it is necessary, in executing this duty, that the municipal authorities adopt measures to fight against noise;

On the proposal of the College of Burgomaster and Aldermen;

After discussion in the Council,

With . . . votes in favour, . . . votes against and . . . abstentions,

orders:

Section I—General Disposition

Article 1.—All unnecessary noise and disturbance, whether by day or at night, of a nature to disrupt the peace and quiet of the residents are prohibited.

Section II—Specific Dispositions

Article 2.—On public ways and in open air public places,

(a) the use of:

amplifiers;

loudspeakers;

music instruments;

other sound apparatus;

(b) the firing of:

firearms;

fireworks;

squibs.

are prohibited.

Article 3.—On public ways, in as far as they are of a nature to disrupt the peace and quiet of the residents, all noises caused by:

(a) the loading, unloading or manipulating of:

engines;

materials;

objects;

(b) the execution of works.

are prohibited.

Article 4.—Besides the prohibitions on public ways and in open air public places, every use of the apparatus or instruments enumerated under article 2, a, in as far as their emission is destined to be heard on the public ways, is also prohibited.

Article 5.—The operators

of dancing halls;

of variety halls;

of theatres;

and, more generally, of all establishments accessible to the public

are obliged to assure that the noise which is produced inside these establishments is not of a nature to disrupt the peace and quiet of the residents.

Article 6.—The obligation formulated under article 5 is applicable as well to:

(a) the organisers of public or private meetings;
(b) the operators of places where such meetings are held.

Article 7.—The burgomaster can derogate from the prohibitions formulated in articles 2 to 4.

Section III—Penal dispositions

Article 8.—Infringements of the present police regulation are punished with imprisonment of from one day to seven days, and with a fine of from one franc to twenty-five francs, or with one of these punishments only.

Notes

1. See article 4.
2. See the Royal Decree of 2 April 1974 relating to the conditions and methods of approval for laboratories and organisations in charge of testing and controlling equipment and devices in the fight against noise, *Moniteur belge*, 30 April 1974, amended by the Royal Decree of 15 April 1977, *Moniteur belge*, 22 June 1977.
3. See note 2.
4. To date, this royal decree does not yet exist.
5. To be multiplied by forty.
6. *Moniteur belge*, 13 July 1976; this royal decree repeals the Royal Decrees of 27 March 1974 and 16 April 1974.
7. See 4.3.
8. This 'Consultation Commission' is composed of twenty-one members, as follows:

    (1) a president
    (2) a delegate of the Minister or Secretary of State responsible for the Environment;
    (3) three members from the Ministry for Public Health and the Family;
    (4) three specialists in science subjects relating to the fight against noise;
    (5) three representatives of organisations the purpose of which is the protection of the environment;
    (6) three representatives of automobile associations;
    (7) three representatives of motorcycle associations;
    (8) two members belonging to the Ministry of Public Works, Town Planning Administration and Land Management;
    (9) one member belonging to the Ministry for National Education & Flemish Culture, Services for Flemish Cluture;
    (10) one member belonging to the Ministry for National Education & French Culture, Services for French culture.

A substitute member is designated for each incumbent. The incumbent and substitute members are nominated by the Minister. The incumbent and substitute members under points 8, 9 and 10 are proposed by the Minister or Secretary of State to whom they are subordinate. They are present at meetings in a consultative capacity.

The members of the Commission were nominated by Ministerial Decree of 26 January 1977, *Moniteur belge*, 25 February 1977.

9. *Moniteur belge*, 26 April 1977.

10. Senate, Questions and replies, extraordinary session of 1977, no. 7, p. 185, Question no. 33 from Mr. Van In.
11. L. Verschooten, Het koninklijk besluit van 24 februari 1977 houdende vastelling van geluidsnormen voor muziek in openbare en private inrichtingen, *Rechtsk. Weekbl.*, 1977–1978, 2604.
12. This provision does not encroach upon the requirement of article 657 of the General Regulation for Protection at Work relating to theatres: 'Necessary measures are taken to ensure that the noise arising from the theatre does not inconvenience the neighbours'.
13. L. Vershooten, *op. cit.* 2606.
14. See Test Aankoop, November 1977, p. 27.
15. *Moniteur belge*, 28 June 1977.
16. Ministerial Decree of 31 October 1977 fixing the maximum subsidy for the purchase of sound meters for the provinces, agglomerations of communes and communes and fixing the conditions to which these sound meters must conform, *Moniteur belge*, 29 November 1977.
17. These amounts are applicable provided that the noise is measured under the conditions and according to the measuring method laid down in attachment 1 of the Directive of the Council of the European Communities of 6 February 1970 concerning the harmonisation of the legislation of Member States relating to the permissible noise level and exhaust devices of motor vehicles. Article 40(2) of the General Regulation provides for an alternative method. For new vehicles, see 4.3.4.
18. Royal Decree of 10 October 1974 acting as a General Regulation on the technical conditions with which motorcycles and autocycles must comply, as well as their trailers, *Moniteur belge*, 15 November 1974.
19. *Moniteur belge*, 24 December 1976.
20. 70/157/EEC; amended by the Directive 73/350/EEC of 7 November 1973. See also Directive 70/388/EEC of 27 July 1970 (acoustic warning devices of motor vehicles) and attachment VI of Directive 74/151/EEC of 4 March 1974 (sound level of agricultural or forestry tractors).
21. *Moniteur belge*, 1 May 1981.
22. *Moniteur belge*, 26–27 July 1937, 29 September 1967 and 15 August 1973. The Law of 27 June 1937 came into force on 31 March 1954.
23. Where necessary, the Law of 18 July 1973 also provides the King with powers with the aim of preventing and combating noise.
24. The standards adopted are those in the attachment to the Convention regarding international civil aviation, signed in Chicago on 7 December 1944. The noise levels engendered by an aeroplane are expressed in EPNdB (Effective Perceived Noise Decibel) in accordance with the standard NBN-SO1-301 'Acoustic—method of evaluation of noise of an aeroplane', published by the Belgian Institute of Standards. It is in fact recognised that the calculation must not only include the intensity of the noise but also the duration of the loudest noise, the height of the tone of the noise and the breadth of the noise spectrum. Only by using EPNdB units can all these elements be taken into account; see H. Mynchke and H. Vandenberghe, Geluid en Geluidshinder, 1975, p. 156.
25. *Moniteur belge*, 11 June 1980 and erratum, 24 June 1980. In its judgement no. 20.915 of 29 January 1981, s.p.r.l. *'Boa Boboli Air Freight'* v. *Minister of Transport*, the Council of State had observed that the Ministerial Decree of 2 May 1975 was to he considered as illegal, because article 19 of the Royal Decree of 15 March 1954 did not empower the Minister of Transport to fix noise standards for subsonic aeroplanes.
26. Executive Decrees—Royal Decree of 11 June 1954: the Royal castles of Laeken and Ciergnon; Royal Decree of 14 April 1958: the Brussels agglomeration.
27. *Moniteur belge*, 4 August 1973.
28. See also the Ministerial Decree of 2 May 1972 fixing the technical measures regarding

the operation of aircraft with a total maximum permitted weight of less than 5,700 kg, used for commercial air transport, *Moniteur belge*, 3 June 1972.

29. Administration of air routes, Belgium, NOTAM 1–2, 2 January 1975 and AgA 2–3–3, 2 April 1975.
30. *Moniteur belge*, 26 March 1954.
31. *Moniteur belge*, 27 September 1966.
32. On this subject, see H. Myncke and H. Vandenberghe, *op. cit.* 141–142. La convention de Rome du 7 Octobre 1962 relative aux dommages causés par les aéronefs aux tiers à la surface, Paris, 1955; A. Clerens, Burgerlijke aansprakelijkheid inzake schaed ann derden op het aardoppervlak toegebracht door luchtvaartuigen. De verdragen van Rome van 1933 en 1952, *Rechtskundig Weekblad*, 1958–1959, 8 March 1959, 377; M. Litvine, La responsibilité pour dommages dérivant du bruit et des detonnations ballistiques provoqués par les aeronefs, Rapport Belge au Congrès International de Droit Comparé, 1970.
33. *Moniteur belge*.
34. *Moniteur belge*, 13 October 1966; see also the Ministerial Decree of 15 March 1967, *Moniteur belge*, 28 April 1967.
35. Royal Decree of 28 March 1969, amended by Royal Decree of 10 July 1973, *Moniteur belge*, 4 and 24 April 1969 and 23 August 1973.
36. Doc. parl. Ch. 1971–1972, no. 192–4, p. 8.
37. This article 59 states that 'the King can issue general regulations on buildings containing necessary provisions to ensure:

(1) the cleanliness, solidity and beauty of constructions, installations and their surroundings, as well as their safety and in particular their protection against fire and flood;

(2) the conservation, cleanliness, safety, good condition and beauty of the public highways, their access and surroundings;

(3) the servicing of buildings by equipment of general interest, with particular regard to the distribution of water, gas, electricity, heating, telecommunications and the removal of refuse;

(4) the comfort of individuals staying in tourist resorts, in particular by the prevention of noise, dust and emanations accompanying the execution of works and their prohibition during certain hours and certain days. These regulations can concern constructions and installations above and below ground, signs, publicity devices, aerials, pipelines, enclosures, depots, plantations, alterations to the landscape, the planning of car parks off the public highway.

These regulations must not encroach upon provisions already imposed by the laws and general regulations relating to public highways. They apply throughout the national territory to that part of the territory and in particular to that agglomeration specified and for which they fix the limits, or even to those categories of communes which they determine.

38. Cass. 3 October 1960, *Pas.* 1961, I, 127.
39. Cass. 29 June 1959, *Pas.* 1959, I, 110.
40. Cass. 28 October 1963, *Pas,* 1964, I. 223.
41. E. Benearts, Competance communale en matière de lutte contre le bruit, in *Convention nationale sur la qualité de la vie*, Liège, 1979, p. 54.
42. Cass. 19 October 1937, *Pas.* 1939, I, 301; Council of State, 28 May 1963, no. 10.044, Goethals.

5
The Protection of Surface Waters

5.1 INTRODUCTION[1]

5.1.1

Among the first provisions issued by the legislator in Belgium for the purpose of ensuring the protection of surface waters against pollution should be mentioned:

certain provisions of the Law of 7 May 1877 on the policing of non-navigable water courses (this law was repealed by the Law of 28 December 1967);

the police regulations on State managed navigable waters;

above all, article 90 of the Rural Code of 7 October 1886 which punishes 'those who throw substances of a nature to destroy the fish into a canal, a lake, a fish-pond or a reservoir'[2] by a fine of from 15 to 25 francs and imprisonment of from 1 to 7 days, or by one of these punishments alone.

A study commission was set up by the Ministerial Decree of 21 December 1900, consisting of eight members, to carry out an investigation into the causes of the growing pollution of surface waters. Its report was never followed up.

In 1929 a new commission for the purpose of making proposals to government was constituted by royal decree; this consisted of delegates from various departments concerned with the problem of water pollution. In its report of 1931, this commission advocated the setting up of a new service to encourage the study, scientific research and promotion of the purification of surface waters. As a result, the Royal Decree of 22

March 1934 constituted a Service for the Treatment of Wastewaters under the Ministry of Public Works; subsequently, in 1939, this service was linked to the Ministry of Public Health.

In 1948 the Treatment Service presented a draft law on the protection of water against pollution. This draft law was unanimously adopted by the Senate in 1949 and approved by a large majority in the Chamber in 1950. The Law of 11 March 1950 on the protection of water against pollution was published in the *Moniteur belge* on 27 April 1950; it was amended twice, first by the Law of 1 July 1955 (*Moniteur belge*, 14 August 1955) and then by the Law of 2 July 1956 (*Moniteur belge*, 8 August 1956).

5.1.2

The Law of 11 March 1950 was aimed at the protection of maritime waters, navigable waters, non-navigable waters, water courses inside polders and drainage canals and all waters on public property in general.

The law prohibited pollution by the throwing or depositing of objects or matter, or by allowing liquids to run into these waters. However, since it was not possible to simply forbid the discharge of all wastewaters, a system of prior authorisation was instituted in order to impose quality standards.

The law made a distinction between three categories of discharge, according to the nature of the place of discharge and of the water discharged, and a classification into four classes of fresh surface waters, according to the predominant use made:

class 1: water used or intended for drinking water supply;

class 2: water used for fishing and animal watering;

class 3: water used mainly for industrial purposes;

class 4: gutters, ditches and aqueducts of the public highways.

5.1.3

Although the Law of 11 March 1950 was considered progressive in its time, 'in practice, its application resulted in a checkmate'.[3]

This unsatisfactory result arose from a variety of reasons. Essential implementation measures were only taken very late with the result that for several years the law remained inoperative; in this field the communes were entrusted with a responsibility far beyond their abilities; useful anti-pollution methods contained in the law were not applied; in addition, an efficient control was almost impossible and the sanctions far too small.

5.2 THE LAW OF 26 MARCH 1971 ON THE PROTECTION OF SURFACE WATERS AGAINST POLLUTION

5.2.1 Basic principles

The protection of surface waters against pollution is currently organised by the Law of 26 March 1971.[4]

It is forbidden to throw or deposit objects or matter, to introduce gas, or to allow polluted or polluting liquids to run into the public water network or coastal waters. The depositing of solid or liquid matter in a place where natural seepage into the above waters could occur is also forbidden.

However, the authorities in charge can authorise the discharge of waste-waters which the law makes subject to authorisation, whatever their level of pollution. The discharge authorisation fixes the conditions to which the discharge must conform. The authorisation can be suspended or withdrawn if these conditions are not respected and the conditions imposed can be modified at any time.

With regard to the previous Law of 11 March 1950, the main innovation consists of the creation of three wastewater purification companies with exclusive responsibility for the fight against water pollution. The Belgian legislator was inspired on this point by French legislation (Agences Financières de Bassins Hydrographiques; Law of 16 December 1964) and English legislation (Water Authorities: Water Resources Act 1963).

The King lays down general regulations relating to sewers—that is, 'all public water networks constructed either in the form of underground pipes, gutters or open ditches used for the collection of wastewaters'—and to the discharge of wastewaters.

In addition, on the advice of the Council of Ministers, the King can control the manufacture, importation, sale and use of products which, if after use they end up in discharges into the sewers or surface waters, are liable to pollute the surface waters or to hinder the self cleansing properties of the water, or to damage the functioning of the wastewater treatment plants operated by the treatment companies created under the law.

5.2.2 Field of application

The law organises the protection of waters in the public hydrographic network and coastal waters against pollution.

The waters of the public hydrographic network include navigable waters and those classed as such, non-navigable water courses and permanent or intermittent water courses, as well as all running and stagnant waters on public lands in general.

Coastal waters include the territorial waters, that is the waters of the coastal sea strip of a width of 3 geographical miles at the rate of sixty miles by degree of latitude from the low tide mark.

Pollution is taken to mean any deposit resulting directly or indirectly from human activities, of matter liable to alter the composition or condition of the water in such a way that it is no longer suitable or is less suitable for the uses to which it could be put, or that it damages the environment because of its appearance or odour.

5.3 ADMINISTRATIVE STRUCTURE

5.3.1 The legal regime

The law envisaged the creation of three wastewater treatment companies:[5]

(1) the company for the waters of the coastal basin;
(2) the company for the waters of the Scheldt basin;
(3) the company for the waters of the Meuse, Seine and Rhine basins.

These treatment companies are public companies and are responsible, each in their own area, for planning, control and research—in short, for the qualitative management of their hydrographic basin. They have an exclusive responsibility on the subject.

5.3.1.1 COMPOSITION OF THE TREATMENT COMPANIES

There are three categories of organisation which must be included as part of these companies:

(1) provinces in which part or all of the area of responsibility of the company lies;
(2) any public organisation which abstracts and distributes water, the offtake of which is located within the area of responsibility of the company;
(3) private businesses with a level of pollution in their discharge waters in excess of the minimum fixed by the King[6] and which, since they do not wish to undertake the treatment operation themselves, entrust it to a treatment plant belonging to the company.

5.3.1.2 INTERNAL ORGANISATION

Each treatment company consists of the following: a general assembly, an administrative council and a director.

The *general assembly* is composed of:

(1) Representatives designated by the deputed states of each of the provinces of the area over which the company's responsibility extends, at the rate of one representative per fifty thousand inhabitants of the province residing within the area in question; each of these representatitves has 50 votes.
(2) A representative of each public organisation which abstracts surface water for distribution, the offtake of which is located within the area of responsibility of the company; each representative has one vote per 350,000 cubic metres of surface water abstracted during the preceding exercise, but their votes cannot exceed 15% of the total.
(3) A representative of each business with a pollution level in its discharge exceeding the minimum fixed by the King, which has entrusted its effluent treatment to a plant belonging to the company; each representative has one vote per sector of a thousand pollution charge units (see 5.3.1.4), but their votes cannot exceed 35% of the total votes.

The *administrative council* is composed of nine members elected by the general assembly as follows:

(1) The representatives of the provinces elect five members from their midst.
(2) The representatives of the public organisations mentioned elect one member from their midst.

(3) The representatives of the businesses mentioned elect three members from their midst.

The representatives of the provinces therefore hold the majority both in the general assembly and the administrative council.

Each treatment company is managed by a *director* who must hold an academic level engineering diploma or doctor of science degree; he is appointed or dismissed by the King on the proposal of the administrative council.

5.3.1.3 RESPONSIBILITIES

As already mentioned, the companies are responsible, each in its own area, for the qualitative management of the hydrographic basin.

In particular their task is:

to establish and ensure the implementation of treatment programmes for wastewaters from the public sewers or where effluent treatment is entrusted to them by private businesses;[7]

to exercise a control over the discharges which are subject to authorisation;

to search for any eventual source of water pollution;

to give opinions on the subject of the measures to be taken in order to protect the waters in their area from pollution.

With regard to treatment programmes, the companies must first take over, manage and improve existing treatment plants belonging to public administrations (chiefly the communes), organisations with public interests or intercommunal associations. In addition, the companies must establish plans for new plants, proceed with the implementation of these plans and ensure their operation and maintenance.

Finally, in exceptional circumstances and with authorisation from the Minister of Public Health, the treatment companies can plan, construct, operate and maintain wastewater treatment plants on behalf of third parties.

5.3.1.4 FINANCIAL MEANS

The treatment companies obtain their finance through:

(1) capital subscription from their associates;
(2) State subsidies;

(3) contracted loans;
(4) contributions from the provinces and private businesses linked to public sewers or to a collector belonging to the company;
(5) the development or sale of treated water or any other matter collected in the course of the treatment process.

The company capital is formed through obligatory subscription when the company is constituted or new members are admitted,[8] according to the following proportions:

(1) the provinces in proportion to the number of their inhabitants residing within the area of responsibility of the company;
(2) public organisations using the surface water, in proportion to the volume of water used, expressed in cubic metres per annum, divided by 350;
(3) the business mentioned above (in 5.3.1.1) in proportion to the degree of pollution of their discharges, expressed in pollution charge units.[9]

The State intervenes financially in *investment* expenditure incurred by the treatment companies, either in the form of capital grants when works are carried out, or in the form of participation in interest charges and loan amortisements contracted by the companies for the purpose of financing their works.[10]

The *management* and *operating* costs are exclusively covered by contributions:

(a) from the associated provinces;
(b) from the businesses whose wastewaters have a pollutant load exceeding seven units for each category of business, and which are treated in an installation belonging to the treatment company;
(c) from the businesses whose wastewaters have a pollutant load which exceeds minima established by the King for each category of business, and which are treated in their own installations.

The provinces are authorised to levy an annual tax in order to recover a part of their contribution towards the management and operating expenses.

The State also intervenes in the cost of *monitoring* carried out by the treatment company on discharge of wastewaters, along lines to be determined by royal decree.

5.3.1.5 DISPUTES

Disputes arising between treatment companies and an organisation or business on the subject of sums due for capital subscription and subsequent increases or for contributions or interest on delayed payments come under civil jurisdiction.

Any action by the treatment company relating to these disputes must be taken within five years from the final date when payment was due.

5.3.2 The present situation

As already mentioned,[11] Belgium is changing into a State 'which will henceforward no longer be centralised but in which certain matters formerly conferred on the central power, will from this time come under the jurisdiction of federal type institutions'.[12] This is true with regard to the treatment of wastewaters.

Since the area limits of two of the three treatment companies do not coincide with the geographic definition of political regions, only the treatment company for the waters of the coastal basin, whose area of responsibility is located entirely in the Flanders region, has been consitituted and functions satisfactorily.[13]

The companies for the waters of the Scheldt basin and of the Meuse, Seine and Rhine basins have not yet been constituted.

With regard to the *Flemish Region,* a Decree of 23 December 1980 of the Flemish Council has amended the Law of 26 March 1971 on the protection of surface waters against pollution, by introducing specific provisions for the Flemish Region.[14]

Implementing this decree, a Royal Decree of 9 March 1981 has created a 'Flemish Water Treatment Company', the area of which covers the Flemish Region, with the exception of those administrative districts covered by the purification company for the waters of the coastal basin.[15] A Royal Decree of 17 April 1981 fixes its statutes.[16]

The Flemish Company started its activities on 28 October 1981.[17]

A Royal Decree of 13 December 1977 controls the treatment of wastewaters in the *Walloon Region.*[18]

The Minister responsible for the treatment of wastewaters in the Walloon Region can:

(1) conclude private contractual agreements with intercommunal associations, by which these associations are entrusted with the responsibility to plan, carry out and manage works for the collection and treatment of wastewaters in the public sewers and the public

drainage network, under conditions laid down in the convention in question.

(2) entrust the management and operation of the installations constructed, or to be constructed, under the conditions determined in the convention.

In fact, the treatment of domestic effluents has been entrusted to nine intercommunal associations: I.D.E.A (Mons), I.E.G.S.P. (Charleroi), Intersud (Thuin), I.N.A.S.E.P. (Namur), A.I.V.E. (Luxembourg), I.B.W. (the Walloon part of the Province of Brabant), S.I.D.E.H.O. (Tournai) and A.I.D. and Inter-cours d'eau (Liège).[19]

The efficiency of such a division of effort may be called into question. It has been said in this connection: 'In short, neither good intentions nor responsible ministers have ever been missing. But by no means has there been efficiency. The treatment of effluent in the Walloon Region is a vehicle that succeeded in starting. But it has not yet succeeded in changing up to second gear'.[20]

5.4 THE AUTHORISATION PROCEDURE FOR THE DISCHARGE OF WASTEWATERS

5.4.1 Authorities responsible

A distinction is made between 'normal domestic' wastewater and 'other than normal domestic' wastewater.[21]

The authorisation to discharge normal domestic wastewater into the public sewers is granted by the College of Burgomaster and Aldermen.

Authorisations to discharge other wastewaters into the public sewers as well as authorisations for all discharges of wastewaters into surface waters are granted by the director of the treatment company responsible for the receiving waters.

Until the treatment company directors have been appointed, these authorisations will be granted by the official in charge of the 'wastewater treatment service' of the Ministry of Public Health.

The authorisation to discharge into coastal waters, navigable waters or those classed as such, as well as non-navigable waters belonging to the State, is only granted with the approval of the public authority, the organisation or concessionaire responsible for the policing or management of the waters in question.

5.4.2 Procedure

5.4.2.1

All authorisation decisions made by the director of a treatment company are notified within eight days to the Minister of Public Health.

During a period of sixty days from receipt of the notification the Minister can cancel or modify the decision. Once this period has expired, the decision becomes definitive.

A copy of the Minister's decision is sent to the director of the treatment company which granted, cancelled or modified the decision.

An administrative appeal against refusal is open to the applicant but this appeal has no suspensive effect.

The appeal must be sent by registered letter within ten days of notification of the decision to the Minister of Public Health and the Environment. Within three months of receiving the appeal the King takes a decision on the justified decree proposed by the Minister, having consulted certain other authorities.[22]

5.4.2.2

Any change in quantity or composition of an effluent is subject to a new discharge authorisation. As is the case with regard to the control of dangerous establishments, the authorisation can be suspended or withdrawn in the event of non-observation of the authorisation conditions; similarly, the conditions imposed can be modified at any time.

The beneficiary of an authorisation which is the subject of such a decision can make an administrative appeal against that decision, as described in 5.4.2.1.

5.4.2.3

In addition to this administrative appeal, it should be remembered that, if necessary, appeal can be made to the Council of State.

5.4.3 Conditions of discharge

5.4.3.1

The conditions of discharge are fixed in accordance with the provisions of the Royal Decree of 3 August 1976 'acting as a General Regulation relating to the discharge of wastewaters'.

5.4.3.2

First, the General Regulation stipulates that it is forbidden to throw or pour either solid wastes which have previously been mechanically pulverised, or waters containing such matter, into the ordinary surface waters, the public sewers or artificial rainwater drainage channels.

5.4.3.3

Discharge authorisations for wastewaters are set up in such a way that conditions imposed on the coastal basin and the basins of the Scheldt, Meuse, Seine and Rhine are identical where the circumstances are identical. These authorisations—independent of the quality objectives recommended by the General Regulation—will in all cases respect the quality objectives imposed by the directives of the European Communities and those contained in the international treaties and conventions to which Belgium is a signatory.

5.4.3.4

The General Regulation makes a distinction according to whether the wastewaters are discharged:

(1) into ordinary surface waters;
(2) into public sewers;
(3) into artificial rainwater drainage channels.

5.4.3.5

In the first two cases, they must respect:

(1) *general* conditions formulated in the General Regulation; these conditions refer to the quality of water discharged (limit values for

pathogenic germs, pH, 5-day Biochemical Oxygen Demand at 20°C, temperature, content of certain substances, etc.);

(2) *sectoral* conditions fixed by royal decree for all businesses in the same sector or sub-sector.[23]

These sectorial conditions can supplement the general conditions, replace them by being more severe, or act as a waiver—only temporarily—in being less severe if it appears that in a sector or sub-sector no commercialised purification process is available to meet the general conditions imposed. In this case a delay is granted for a fixed period at the end of which the conditions must either be confirmed or abolished, depending on whether appropriate scientific and technical progress has occurred in the meantime.

The sectoral conditions consist in particular of the following:

(1) Maximum quantities of pollutants per unit produced. In this case the discharge authorisation specifies the concentration limits admissable in the quantities of water discharged and the production, in each individual case.

(2) Maximum quantities of water which can be discharged per unit produced.

To date, sectoral conditions have been fixed for forty-one sectors.[24]

(3) *special* conditions fixed in the authorisation.[25]

As is the case for sectoral conditions with regard to general conditions, special conditions can complete the general and sectoral conditions or replace them in being more or less severe, but in the latter case, the waiver granted should either be accompanied 'by a fixed period at the end of which the conditions must be either confirmed or partially or totally abolished, depending on the scientific and technological progress achieved in the meantime' (article 10(4) of the General Regulation), with regard to discharges into the ordinary surface waters; alternatively, the waiver can be the subject of an agreement between the discharger and the authority operating the plant treating the industrial effluent where a discharge into the public sewers is involved.

Finally, the discharge authorisation can:

(1) require the discharge of different types of effluent to be carried out separately;

(2) impose requirements:
   (a) to limit the risk of accidental discharge of wastewaters or dangerous substances;
   (b) to ensure that the authority responsible for authorisaton of the discharge is informed immediately of such a discharge;

(3) specify:
- (a) the maximum quantity of wastewater that can be discharged daily, as well as the maximum instantaneous flows;
- (b) that the discharge should be emitted through a device:
  - (i) indicating either at the moment of measurement or continuously the instantaneous flow or certain other characteristics of the discharge;
  - (ii) indicating the daily quantity of discharge;
  - (iii) facilitating samples to be taken of the discharge;
- (c) that an inspection chamber to facilitate the taking of samples from the discharge should be installed in all outflow conduits.

5.4.3.6

No discharge of wastewater into an artificial rainwater drainage channel from a public highway is permitted if the public highway is provided with public sewers.

When the public highway is not provided with public sewers, the discharge of normal domestic wastewater[26] into an artificial rainwater drainage channel can be authorised under the following conditions:

(1) The waters discharged must be effectively treated by a device to eliminate greasy substances, floating and decantable matter.
(2) The waters discharged must not contain more than 5 mg/l of petroleum ether extract.
(3) The waters discharged must not emit or cause the emission of noxious odours.

The discharge of wastewaters other than normal domestic effluent into artifical rainwater drainage channels is forbidden.

5.4.3.7

The conditions contained in an authorisation must be respected within forty months at the latest, from approval of the authorisation.

5.5 CONTROL AND SANCTIONS

5.5.1 Powers of the agents and officials appointed by the Minister

The Law of 26 March 1971 provides for the appointment by the Minister of Public Health of specially qualified agents to monitor the application of the law and its implementing decrees; these in no way detract from the powers of the judiciary police officers.[27]

These agents ascertain infringements of the laws and devices by written reports which act as evidence in the absence of proof to the contrary. A copy of the report is sent to the person who committed the infringement within three days from ascertainment of the infringement.

They have the right to enter establishments and installations at all times—excluding residential premises—when they have reason to believe that an infringement of the law or decrees relating to the protection of surface waters against pollution is being committed.

If sufficient indications exist to suppose that an infringement of this kind is being committed in a residential premises, a house visit can be made by two agents acting on authorisation from the police tribunal judge.

The agents appointed by the Minister can request the assistance of the communal authorities in carrying out their duties.

5.5.2 Monitoring the discharge of wastewaters

The appointed agents can take samples of waters discharged and of receiving waters and have them analysed by a State laboratory or a laboratory approved for this purpose by the Minister of Public Health.[28]

If a discharge into ordinary surface waters is involved, at least one sample of the discharged water is taken; then, at least one sample will also be taken from the receiving waters above the discharge point. Monitoring of discharges into public sewers or artificial rainwater drainage channels is carried out through sampling of the discharged water. Each sample is made up of two equal parts of at least three litres.

The samples, whether for analysis or checking, are handed over or sent within twenty-four hours of being taken, to the State laboratory or approved laboratory in charge of the official analysis.

A written report is prepared of the sampling operation and is sent to the following within three working days from the date when the sample was taken:

(1) the authority responsible for the approval of the discharge authorisation;
(2) the holder of the discharge authorisation;
(3) the official directing the wastewater treatment service of the Ministry of Public Health and the Family;

(4)   to the head of the laboratory carrying out the official analysis of the sample taken.

During the five working days following the sampling operation, one part of the sample is held at the disposal of the holder of the discharge authorisation for him to analyse.

Monitoring of discharges can also be carried out by means of equipment approved by the Institute of Hygiene and Epidemiology, which can be used to determine the characteristics of the discharge on the spot.

In this case, either the authorisation holder for the discharge under examination, or a representative from an approved laboratory appointed by him, must be present when the control is carried out.

5.5.3 Emergency measures

The director of a treatment company as well as the appointed agents can request the communal authorities to take emergency measures as a result of the existence or imminent danger of serious pollution.

In the event of inaction on the part of the communal authorities or when the polluted waters are liable to constitute an imminent danger or to cause serious damage to the population, the director of a treatment company or the appointed agents can make the necessary requisitions under whatever provisions they consider appropriate and on their own responsibility.[29] They immediately inform the Minister of Public Health and the governor of the province.

The implementation of these requisitions is ensured by the intervention of either the governor of the province or the district commissioner of the area. They can take all the appropriate measures at once on the spot and can make use of the police force and the gendarmerie for this purpose.

All such measures cease to have effect after a period of thirty days if, during that time, they have not been confirmed by the Minister of Public Health, the users having been previously heard or called upon to reply.

5.5.4 Infringements

The definition of infringements envisaged by the law and the punishments applicable are the subject of article 4, as follows:

(1) Without prejudice to the punishments under the Penal Code, those who contravene the provisions of the law or implementing decrees and regulations as listed below are punishable by imprisonment of from eight days to six months and by a fine of from 26 to 5,000 francs, or by one of these punishments alone:

(a) those who, in infringement of article 2, throw or deposit objects or matter, or allow liquids containing harmful matter or substances to run into the waters covered by article 1, or who introduce gas into these waters, as well as those who, on their orders or by negligence, cause an action of that kind, or cause solid or liquid matter to be deposited in a place where it can be introduced by a natural phenomenon into the waters under article 1;

(b) those who, in infringement of article 5, discharge used waters into the waters under article 1 or into the public sewers without prior authorisation or without conforming to the conditions imposed in each authorisation;

(c) those who voluntarily destroy or damage treatment installations or hinder their operation in any way;

(d) those who obstruct the implementation of the control, monitoring or investigation tasks of personnel under articles 36 and 37.

(2) The judge can prohibit for a fixed period the use or operation of the installation or equipment which has caused the infringement.

(3) The punishments are doubled if a further infringement is committed within two years from a previous judgement for one of the infringements under this article and carrying the force of a judgement.

(4) Legal persons are civilly responsible for judgements of damaged interests, fines and costs inflicted on their organisations or employees who committed infringements while carrying out their duties.

5.5.5

When a communal authority discharges sewage into the surface waters without authorisation or without conforming to the authorisation conditions, the governor or the deputed states of the provincial council, after two notifications, can summon one or more commissioners at the personal expense of the communal authorities to implement the notifications and carry out the prescribed measures.

It should, however, be noted that the discharge of treated urban wastewater is subject to specific standards. Indeed these waters are subject to the same discharge conditions as those provided for in the treatment plant project approved by the minister with regional responsibility for wastewater treatment matters; this project must conform with the general operating conditions fixed by the minister with national responsibility for the treatment of wastewaters (article 16, Royal Decree of 3 August 1976, acting as a General Regulation).

5.5.6

When the director of a treatment company ascertains that, in spite of legal proceedings against infringements, the waters continue to be polluted in some way, he makes a report and proposes appropriate measures to the Minister of Public Health.

The Minister orders the necessary measures to put an end to the situation, in particular, suspension of the discharge authorisation and prohibition on use of the installations or equipment which may be the cause of the pollution.

He can have installations and equipment sealed off by officials appointed by him.

The measures ordered by the Minister of Public Health are implemented at the costs, risks and danger of the party involved.

5.6 FINANCIAL INTERVENTION BY THE STATE

5.6.1

As a result of article 33 of the Law of 29 March 1971, the State intervenes in *investment* expenditure made by industrial businesses for the purpose of fulfilling legal requirements, in two cases:

(1) When the business, at the time of its establishment in a specified area, and for valid reasons, does not use one of the treatment company's plants to treat its wastewaters.
(2) When the discharge authorisation granted to an already existing business requires its wastewaters to undergo special treatment which would involve additional investment.

5.6.2

The second type of intervention is currently controlled by the Royal Decree of 9 April 1975 'relating to the intervention of the State in the additional investment to which an already existing business is committed for the special treatment of its used waters'.[30]

Special treatment is taken to mean either treatment of wastewaters in such a way as to respect the conditions imposed, or complete elimination of discharges of wastewaters, or adaptation of production processes with the same effect as the two cases mentioned above.

5.6.3

Businesses can only benefit from this State intervention for manufacturing installations which were operating on 1 January 1975.

The plans for installations for which State intervention is requested must be studied by a draftsman approved by the minister responsible for the environment, the choice of whom is approved by the Minister or Secretary of State with regional responsibility for the protection of surface waters against pollution.[31]

Intervention by the State can be approved not only for expenditure relating to the special treatment of used waters, but also for expenditure relating to all the works necessitated by the special treatment.

Investment expenditure which serves as the basis for the calculation of State intervention will also include the costs of studies fixed at a maximum of 7% of these costs, as well as the non-deductible part of the value added tax relating to business and study contracts.

However, the expenditure for eventual land purchases required for works to be carried out is not taken into consideration for the subsidy.

The subsidy is only granted for the portion of investment expenditure in excess of 200,000 francs.

5.6.4

State intervention diminishes progressively. The intervention rate is fixed at:

45% if the State budgetary commitment is made between 1 January 1975 and 31 December 1976;

30% if the State budgetary commitment is made between 1 January 1977 and 31 December 1978;

15% if the State budgetary commitment is made after 1 January 1979.

Where appropriate, the intervention is decreased by the amount of any other State subsidy granted for the same project.

5.6.5

In fact, the available credit has reached the following amounts (in millions of Belgian francs):

	Intervention rate	Flanders Region	Walloon Region	Brussels	Total
1975	45%				150.3
1976	45%	1,544	517.2	—	2,061.2
1977	30%	639.4	41.9	—	661.3
1978	30%	826	212	29	1,067
1979	15%	500	500	80	1,080

5.6.6

State intervention is granted either:

(1) in the form of direct payments to the business applying; or
(2) in the form of an undertaking by the State to pay the interest and amortisation charges on the loans contracted by the business, through credit organisations approved by the Minister of Finance who makes an agreement with them for this purpose.

5.6.7

To obtain State intervention, the business must formally agree:

(1) to construct the installations not yet completed, within the period specified in the discharge authorisation;

(2) to keep the installation operating in such a manner that the discharge conditions imposed in the authorisation which forms the basis for State intervention, are respected at all times and that the sludge arising from the process is treated effectively;

(3) to ensure the operation of the installation and to be responsible for all the resulting expenses.

If it is proven that the business which has obtained State intervention is not respecting the conditions and agreements linked with that intervention and that this failure results from a defect in construction or a permanent malfunction or lack of proper maintenance or even a failure to adapt the installation to a change in the production of the pollutants, the State can require the repayment of the grant.

5.6.8

With regard to the *operating* costs of installations belonging to industrial businesses, article 35(1) of the Law of 29 March 1971 states:

> At the request of all interested businesses, each treatment company in the area in which the businesses are located will bear all or part of the operating costs of installations for the treatment of their wastewaters—if these cannot be treated in installations belonging to those companies.

A royal decree can be adopted to ensure equality of financial charging between the various treatment companies in their application of this article 35(1).

5.7 THE PREVENTION OF POLLUTION OF SURFACE WATERS BY USED OILS

As a result of the Royal Decree of 3 October 1975 relating to the prevention of pollution of surface waters by used oils,[32] it is forbidden:

(1) to deposit, throw, pour or allow used oils, whether in a container or not, to run into the waters of the public hydrographic network or coastal waters, or to deposit them in a place from which they could enter the waters described above as a result of a natural phenomenon;

(2) to deposit, throw, pour or allow used oils, whether in a container or not, to run into the public sewers or into artificial rainwater drainage channels;

(3) to collect used oils without the previous approval of the Minister of Public Health;

(4) to deliver or consign used oils to non-approved collectors;

(5) to import or export used oils without previous authorisation from the Minister of Public Health;

(6) to add water or any other substance to used oils before or during their collection.

Retailers of oils and mineral oils are obliged to accept used oils handed to them by their customers in maximum quantities equal to the quantities of new oil or mineral oils sold to them, without requiring any payment.

Infringements of these provisions are investigated, verified and punished according to the Law of 26 March 1971 on the protection of surface waters against pollution.

5.7.1 Approval qualifications for a collector

In order to qualify as an approved collector of used oils the applicant must:

(1) formally agree to collect, within a period not exceeding fifteen days from the date of a collection request, throughout the whole territory specified in the approval act, all used oils arising in more than a minimum quantity specified in the approval act, without requiring payment on the part of those surrendering the oils, except when otherwise specified in the approval act; in this case, the approval act is jointly signed by the Ministers of Public Health and Economic Affairs;

(2) to formally agree not to mix synthetic oils with mineral oils;

(3) to prove that he has available, himself or by means of contracts with third parties, sufficient technical and financial means to guarantee the collection, transport, storage and non-polluting use of used oils in accordance with the existing laws and regulations.

The approval can only be agreed for a territory which includes at least one province.[33] The approval does not confer a collection monopoly for the designated territory.

Approved collectors are obliged to give a certificate, which should be kept for three years, to the person who delivered the used oils to them; in addition, they must also keep an up-to-date register—also to be kept for three years—indicating the daily quantities of used oils collected and the use made of them.

5.7.2 Authorisation to import or export used oils

In order to become an authorised importer or exporter of used oils,[34] the applicant must supply proof that he has available, either himself or through contracts with third parties, sufficient technical and financial means to guarantee the non-polluting transport, storage and use of used oils in accordance with existing laws and regulations.

5.7.3 Procedure

The application for approval as a collector and for authorisation as an importer or exporter of used oils must be sent by registered letter to the Minister of Public Health.

The decision is notified to the applicant by registered letter. The reasons are given in the event of refusal.

The activity of one or more approved collectors in the same territory does not constitute a reason for refusal.

In the event of approval of a collector, an extract of the decision is published in the *Moniteur belge*. This extract mentions the identity of the approved collector as well as the territory for which the approval is granted.

5.7.4

An approval or authorisation can be suspended or withdrawn if their holder does not fulfil the conditions or if he does not respect the regulatory provisions or the particular conditions specified by the minister in charge.

5.8 DETERGENTS

5.8.1

Finally, mention should be made of the Royal Decree of 23 March 1977[35] which prohibits the import, sale and use of detergents if the average biodegradability of the surface agents contained therein is less than 90% for each of the following categories: anionic, cationic, non-ionic and ampholytic.

The use of surface agents with an average biodegradability of at least equivalent to 90% should not, under normal conditions of use, result in damage to the health of man, animals or vegetation.

The admissable tolerance with regard to the biodegradability of anionic surface agents is fixed at 10%.

5.8.2

The European agreement to restrict the use of certain detergents in washing and cleaning products, made in Strasbourg on 16 September 1968 and approved by the Belgian Law of 28 February 1970,[36] states that washing or cleaning products containing one or more synthetic detergents can only be sold on condition that the mixture of detergents in the product in question is at least 80% biodegradable.

Notes

1. See the study by N. De Baenst, Een nieuwe wet op de bescherming van de opperv-laktewateren tegen verontreiniging, *Gemeentekrediet van Belgie,* 97, July 1971, which provides detailed information on the subject of the history of water pollution legislation.
2. The practical extent of this provision is minimal, contrary to the similar provision under article 434 (1) of the French Rural Code, of which M. Despax estimates that 'of all our legislative arsenal, it is assuredly the most used and the most feared weapon' (M. Despax, Droit de l'environnement, Paris 1980, pp. 369–398.
3. J. Constant, La protection pénale de l'environnement en droit Belge, in *Rapports belges au Xe Congrès international de droit comparé,* Brussels, 1978, p. 575.
4. *Moniteur belge,* 1 May 1971.
5. The Royal Decree of 26 July 1972 determines the limits of the areas of responsibility of these companies, *Moniteur belge,* 29 August 1972. The statutes of the companies are fixed by royal decree. The companies are subject to the rules imposed by the Law of 16 March 1954 relating to the control of certain organisations of public interest, on category B establishments.
6. The Royal Decree of 23 January 1974 fixes this minimum at 5,000 pollution charge units (*Moniteur belge,* 15 February 1974).
7. These purification programmes are subject to the approval of the Minister of Public Health (article 10 of the law).
8. Article 15 contains a detailed control of the problems posed by an increase in capital or by the obligation to recalculate the portions relating to each of the associates every five years in the absence of a capital increase.
9. The Royal Decree of 23 January 1974 defined the pollution charge unit as a function of the volume and characteristics of the used water normally discharged by one

person in 24 hours, as follows: 180 litres of used water with a content of suspended matter of 500 mg/l, a 5-day biochemical oxygen demand of 300 mg/l, a chemical oxygen demand of 750 mg/l and a Kjeldahl nitrogen content of 55 mg/l.

The values of conversion coefficients for pollution charge units of used waters arising from industrial or other businesses have been determined by the Ministerial Decree of 15 February 1974, amended by Ministerial Decree on 3 March 1975, 27 May 1977 and 27 December 1977, *Moniteur belge*, 29 March 1974, 29 August 1975, 15 October 1977 and 6 May 1978.

It is intended to determine separately for each individual business the number of pollution charge units, on the basis of analyses of the used water; this more accurate method would be preferable to the provisional system currently in force.

10. The Royal Decree of 20 December 1976, which determines the conditions and methods of State financial intervention in the investment expenditure of the only, at that moment, existing treatment company—that for the waters of the coastal basin— fixes the intervention rate at 100% of the investment sum.
11. See Chapter 1.
12. Chr. Lambotte, Le démembrement de la norme juridique en Belgique, in *Bulletin de l'I.D.E.F.*, 1980, no. 34, p. 6.
13. See the following royal decrees:

Royal Decree of 9 April 1975 fixing the statutes of the treatment company for the waters of the coastal basin created by the Law of 26 March 1971, *M.B.*, 8 May 1975.

Royal Decree of 16 April 1975 fixing the portion of capital for each of the associates of the treatment company for the waters of the coastal basin, *M.B.*, 8 May 1975.

Royal Decree of 23 September 1976 fixing, under the heading of affiliation, the portion of capital for each of the associates of the treatment company for the waters of the coastal basin, *M.B.*, 15 December 1976.

Royal Decree of 12 April 1977 completing Royal Decree of 23 September 1976 fixing, under the heading of affiliation, the portion of capital for each of the associates of the treatment company for the waters of the coastal basin, *M.B.*, 6 September 1977.

Royal Decree of 14 May 1975 fixing the date for the commencement of activities by the treatment company for the waters of the coastal basin, *M.B.*, 15 May 1975. This date was 16 June 1975.

Royal Decree of 16 April 1975 fixing the personnel cadre of the treatment company for the waters of the coastal basin, *M.B.*, 8 August 1975.

Royal Decree of 7 July 1975 fixing the personnel statute for the treatment company for the waters of the coastal basin, *M.B.*, 8 August 1975.

Royal Decree of 9 June 1975 nominating a government commissioner for the treatment company for the waters of the coastal basin, *M.B.*, 19 September 1975.

Royal Decree of 10 June 1975 nominating the members representing the most representative organisations of employers on the Administrative Council of the purification company for the waters of the coastal basin, modified by the Royal Decree of 11 September 1975, *M.B.*, 19 September 1975.

Royal Decree of 20 December 1976 determining the conditions and methods of financial intervention by the State in the investment expenditure of the treatment company for the waters of the coastal basin, *M.B.*, 22 January 1977. During the years 1977 and 1978, the treatment company for the waters of the coastal basin carried out works amounting to a total of 1,503 million francs. These are basically plants, drains and collectors.

14. *Moniteur belge*, 26 February 1981.
15. *Moniteur belge* 3 April 1981.
16. *Moniteur belge* 19 May 1981.
17. See the Royal Decree of 15 October 1981, *Moniteur belge*, 20 October 1981.
18. *Moniteur belge*, 10 October 1979.
19. See Etienne Orban, Les aspects juridiques de l'épuration des eaux usées ... en region Wallonne, *Rev jur éc. urban. environn.* Seres, June 1980, pp. 1–14.
20. Jacques Poncin, Dépolleur les eaux? Parfait, mais la loi est boiteuse, les compétances rares et les entreprises trop pauvres!, *Le Soir*, 13 November 1980, p. 5.
21. These two notions are meticulously defined in article 1 of the Royal Decree of 3 August 1976 acting as a General Regulation relating to discharges of used waters into the ordinary surface waters, into public sewers and into the artificial rainwater drainage channels, *Moniteur belge*, 29 September 1976; err. 11 November 1976. Broadly, 'normal used domestic waters' can be considered as all used waters arising from activities of a domestic nature, or from the cleaning of buildings as well as all used waters arising from limited professional activities—such as businesses employing fewer than seven people or establishments looking after or rearing a fairly large number of animals.
22. See article 2 of the Royal Decree of 23 January 1974 fixing the conditions and terms for the appeal instituted by the Law of 26 March 1971 on the protection of surface waters against pollution, *Moniteur belge*, 15 February 1974.
23. The discharge conditions for electricity power stations are fixed case by case.
24. Abattoirs: Royal Decree of 3 August 1976 (*Mon.*, 29 September; err. 11 November, modified by Royal Decree of 22 April 1977 (*Mon.*, 4 June). Poultry abattoirs: Royal Decree of 22 April 1977 (*Mon.*, 4 June). Starch: Royal Decree of 22 April 1977 (*Mon.*, 4 June). Laundries: Royal Decree of 3 August 1976 (*Mon.*, 29 September). Breweries, maltings, processing and bottling of drinks: Royal Decree of 3 August 1976 (*Mon.*, 29 September; err. 11 November). Quarries, cement factories etc.: Royal Decree of 22 April 1977 (*Mon.*, 4 June). Collieries and associated activities: Royal Decree of 3 August 1976 (*Mon.*, 29 September). Chlorine (industry of): Royal Decree of 22 April 1977 (*Mon.*, 4 June). Cokeries: Royal Decree of 3 August 1976 (*Mon.*, 29 September). Vegetable preserving: Royal Decree of 3 August 1976 (*Mon.*, 29 September; err. 11 November). Distilleries and yeast makers: Royal Decree of 3 August 1976 (*Mon.*, 29 September; err. 11 November). Manure: Royal Decree of 22 April 1977 (*Mon.*, 4 June). Gelatine: Royal Decree of 22 April 1977 (*Mon.*, 4 June). Oils and vegetable and animal fats: Royal Decree of 22 April 1977 (*Mon.*, 4 June). Chlorinated hydrocarbons: Royal Decree of 22 April 1977 (*Mon.*, 4 June). Printing works: Royal Decree of 23 January 1979 (*Mon.*, 24 March). Pharmaceutical industry: Royal Decree of 22 April 1977 (*Mon.*, 4 June). Laboratories: Royal Decree of 22 April 1977 (*Mon.*, 4 June). Woollens (washing): Royal Decree of 3 August 1976 (*Mon.*, 29 September; err. 11 November). Flax (setting of): Royal Decree of 22 April 1977 (*Mon.*, 4 June). Cold mechanical processes and surface treatment of metals: Royal Decree of 3 August 1976 (*Mon.*, 29 September). Non-ferrous metals: Royal Decree of 3 August 1976 (*Mon.*, 29 Setember). Cleaning of casks: Royal Decree of 22 April 1977 (*Mon.*, 4 June). Cleaning of vehicles and river boats used for the transport of liquids: Royal Decree of 22 April 1977 (*Mon.*, 4 June). Paper pulp, paper and cardboard: Royal Decree of 3 August 1976 (*Mon.*, 29 September). Modified by Royal Decree of 22 April (*Mon.*, 4 June). Peroxide (manufacture of): Royal Decree of 22 April 1977 (*Mon.*, 4 June). Petrochemicals and organic chemicals: Royal Decree of 22 April 1977 (*Mon.*, 4 June). Petroleum (refineries): Royal Decree of 3 August 1976 (*Mon.*, 29 September). Fish (preparation of): Royal Decree of 3 August 1976 (*Mon.*, 29 September). Potatoes (treatment of): Royal Decree of 22

April 1977 (*Mon.*, 4 June). Piggeries etc.: Royal Decree of 22 April 1977 (*Mon.*, 4 June; err. 19 July). Soapworks etc.: Royal Decree of 22 April 1977 (*Mon.*, 4 June). Hot iron metallurgy: Royal Decree of 3 August 1976 (*Mon.*, 29 September; err. 11 November). Sugar (industry of) and beet refineries: Royal Decree of 3 August 1976 (*Mon.*, 29 September; err. 11 November). Tanning and leather dressing: Royal Decree of 3 August 1976 (*Mon.*, 29 September). Modified by Royal Decree of 11 April 1978 (*Mon.*, 5 July). Textile (finishing): Royal Decree of 3 August 1976 (*Mon.*, 29 September; err 11 November). Veal (fattening): Royal Decree of 22 April 1977 (*Mon.*, 4 June). Varnish and paint (production of): Royal Decree of 22 April 1977 (*Mon.*, 4 June). Meat (processing): Royal Decree of 22 April 1977 (*Mon.*, 4 June). Viscose (production of): Royal Decree of 3 August 1976 (*Mon.*, 29 September; err. 11 November).

25. With regard to discharges of wastewaters into ordinary surface waters, the particular conditions are those resulting from local conditions which are necessary in order to:

    (a) take account of the directives ot the European Communities relating to the quality of surface waters;

    (b) take account of other international agreements relating to the quality of surface waters;

    (c) take account of the normal functioning of industries and enable new industries to be installed;

    (d) take account of certain quality objectives contained in annex VI of the General Regulation, which should be achieved ten years after publication of the General Regulation, i.e. 1986;

    (e) avoid deterioration in the existing situation of receiving waters or maintain the improved level achieved. With regard to discharges of used waters into the public sewers, particular conditions can be imposed to avoid:

        (i) danger to the maintenance personnel working on sewers and treatment plants;

        (ii) a risk of deterioration or obstruction of canals;

        (iii) any hindrance to the good functioning of compression or treatment plants;

        (iv) a risk of serious pollution of the receiving waters into which the public sewer discharges.

26. With the exception of waters containing matter of faecal origin, for which specific conditions are fixed in article 29 of the General Regulation.

27. The following agents were appointed by Ministerial Decree of 28 April 1975 (*Moniteur belge*, 4 June 1975):

    (1) the directors of the treatment companies and members of company personnel designated by them;

    (2) the director of the Institute of Hygiene and Epidemiology of the Ministry of Public Health and the Family, and the officials designated by him;

    (3) the official directing the Sanitary Engineering Service of the same Ministry and the officials designated by him;

    (4) the agents with the task of monitoring the public administrations, organisations of public interest and concessionary companies responsible for the policing or management of the waters protected by the Law of 26 March 1971;

    (5) the agents and employees of the Water and Forestry Administration who are in charge of policing, monitoring and conserving river fish;

    (6) the officials and technical agents of the Provincial Hygiene Institute of Antwerp, in the province of Antwerp;

    (7) the officials and technical agents of the Provincial Hygiene Institute of Mons, in the province of Hainault.

28. See Ministerial Decree of 25 July 1975 fixing the approval conditions for the test laboratories, *Moniteur belge*, 4 October 1975.

29. When the imminent danger or serious damage results from a collective or voluntary

stoppage of work or from a collective dismissal of personnel, the director of the treatment company will only intervene if the measures taken in application of the Law of 19 August 1948 relating to activities in the public interest in time of peace appear to be inoperative.

30. *Moniteur belge*, 24 June 1975, as amended by Royal Decree of 5 November 1976, *Moniteur belge*, 27 November 1976, and by Royal Decree of 11 August 1977, *Moniteur belge*, 4 October 1977.

31. See the Ministerial Decree of 17 April 1974 approving the project authors of certain works financed or subsidised by the Minister of Public Health and the Family, modified by Ministerial Decree of 31 July 1975, *Moniteur belge*, 15 June 1974 and 5 September 1975.

32. *Moniteur belge*, 25 November 1975. 'Used oils' are taken to mean all mineral or synthetic oils or fats, including the oily residues from tanks, oil and water mixtures and emulsions, whether they have been adulterated as a result of the use made of them, and if they have not been used because they have become too dirty for the use for which they were intended and therefore have no commercial value for their holder.

33. An exception is made for physical or legal persons who, at the date when the Royal Decree of 3 October 1975 came into force—i.e. 15 November 1975—already collected used oils; they were allowed to pursue their activity provided they applied for approval within three months.

34. See article 6 of the Royal Decree of 3 October 1975 which states: 'The physical or legal persons who, at the date of the entry into force of the present decree, collect, import or export used oils can pursue their activity provided they have presented an application for approval or authorisation within three months from that date, and until a decision is given'.

35. *Moniteur belge*, 9 June 1977; see also directives 73/404/EEC and 73/405/EEC of the Council of the European Communities, *Official Journal* 17 December 1973, L347.

36. *Moniteur belge*, 20 November 1970.

6
Groundwaters

6.1 THE LAW OF 26 MARCH 1971 ON THE PROTECTION OF GROUNDWATERS

6.1.1 Introduction

The Law of 26 March 1971 on the protection of groundwaters forms the second part of the new legislation on water protection.

Unlike the law on the protection of surface waters, this is an outline law based upon implementation by decrees which have *not yet been issued*.

6.1.2 Implementation and scope

The implementation of this law is the exclusive responsibility of the Minister of Public Health. It considers groundwaters to include 'all water which does not belong to the hydrographic network, and by assimilation, all water contained in intake pipes and intended for water supply'.

In the interests of public health, the King, in consultation with the Superior Water Distribution Council, can take all necessary measures to protect groundwaters intended for eventual domestic use and water supply purposes.

6.1.3 System of zoning

6.1.3.1

Thus, the King can identify 'catchment zones' and 'protection zones'.

A 'catchment zone' is a geographic area 'in which works or installations are situated intended for the abstraction or storage of groundwaters used, or liable to be used principally in the distribution of potable water'.

A 'protection zone' is a geographic area defined 'for the protection of water in the distribution system and intended for potable supplies, from any risk of alteration'.

In catchment zones and protection zones, the protection of groundwaters is ensured by the operator.

6.1.3.2

The Law of 26 March 1971 contains two principles on procedures:

(1) the identification of a catchment zone or a protection zone will be fixed after a public inquiry, by a royal decree debated by the Council of Ministers;
(2) the Minister of Public Health will decide on any complaints.

Jurisdictional appeal to the Council of State is possible.

Royal implementing decrees still to be issued will determine:

(1) the methods and periods for the public inquiry and, in particular, the procedure for registering and examining complaints relating to the establishment and definition of catchment and protection zones;
(2) the authorities in charge of inquiries and examination of complaints.

6.1.4 Legal conditions and prohibitions in the public interest

6.1.4.1

In catchment and protection zones a royal decree can prohibit, control or make subject to authorisation:

(a) the transport, storage, depositing, discharge, burying, throwing away, pouring and spreading of matter liable to alter groundwaters;

(b) works, installations and activities, as well as any alteration to the soil or subsoil which could constitute a cause or a risk of alteration to groundwaters.

In addition, outside catchment and protection zones, the King can prohibit, control or make subject to authorisation the direct or indirect pouring away or depositing on the ground or into the soil of matter which he declares to be likely to alter the groundwaters.

6.1.4.2

A royal decree—also not yet issued—will determine the methods and fix the period during which authorisation applications must be presented, in implementation of the above decrees mentioned in 6.1.4.1.

The Minister of Public Health will decide on these applications by giving reasons for refusal or specifying the conditions imposed in each individual case.

The decision will only become final after it has been ascertained that the conditions imposed are respected. The methods and periods for this procedure will be determined by the King.

If the conditions are not respected, the Minister can suspend or withdraw the authorisation, giving reasons for his decision, and, where appropriate, he can impose new conditions.

Here, too, there is the possibility of an appeal through the normal administrative contentious channels, that is to the Council of State.

6.1.4.3 EXPROPRIATION

By authorisation from the King, in accordance with the laws on State expropriation, the operator can proceed with the expropriation of immovable property which is essential in achieving the objectives of the present law, in the public interest.

These expropriations can be carried out with the intervention of immovable property acquisition committees acting on behalf of the State, the officials of which are empowered to pass acts, institute proceedings and direct expropriation procedures on behalf of the operator.

6.1.4.4

The direct and material damages suffered by the owner or operator resulting from a legal charge issued in the implementation of the law, in the public interest, are compensated for and the costs of such compensation are borne by the beneficiary of the protection.

This applies to existing works or installations which must undergo alterations or be done away with, as well as to any activities which have to be ceased, reduced or altered.

The right to compensation also exists if duly authorised works or installations are further suppressed or altered, or if activities are to be stopped, reduced or altered as a result of a decision by the authority in charge.

In the event of legal proceedings the compensation fixed by the tribunal can be obtained, notwithstanding any possible appeal.

6.1.4.5 SUPERVISION OF APPLICATION OF THE LAW

The Law of 29 March 1971 provides for the designation of specially skilled agents to supervise the application of the law and its implementing decrees.

The written reports which they prepare are accepted as evidence unless there is proof to the contrary. A copy is sent to those who contravene the provisions within seven days from the ascertainment of such a contravention.

These agents can enter establishments or installations night or day—with the exception of premises for residential use—when they have reason to believe that the present law or its implementing decrees are being infringed.

If there are sufficient indications to presume that an infringement of this kind is being committed in premises for residential use, a house visit can be made by two of these agents, acting on the authorisation of the police tribunal judge.

They can take samples, or cause samples to be taken, of substances liable to make, or presumed to have made, protected waters unsuitable for domestic or water supply purposes.

On the demand of the operator of a water supplies installation, they can also take soil samples if it is presumed that the soil may contain substances of this kind. The analysis of samples is carried out by a State laboratory or by a laboratory approved for this purpose by the Minister of Public Health.[1]

These measures cease to have effect after a period of fifteen days if, during that period, they have not been ratified by the Minister of Public Health or his delegated official, having previously consulted or called the users of the installations or equipment in question.

Ratification decisions are notified without delay, by registered letter, to the users.

Appeal to the King against ratification decisions is open to all interested parties. The King fixes the methods and periods within which this appeal must be made and it has no suspensive effect.

Finally, the designated agents can request the assistance of the communal authorities in carrying out their duties. They can also request these authorities to take any urgent measures necessary when groundwaters become unsuitable for domestic and water supply purposes.

In the event of neglect by the communal authorities or when the slightest delay is liable to result in serious damage to public health, the designated agents themselves, on their own initiative, can take appropriate measures or arrange the necessary requisitions.

They must immediately notify the Minister of Public Health as well as the governor of the province.

The implementation of these provisions is ensured on the spot either by the intervention of the governor of the province or by the commissioner of the district.

These measures cease to have effect after a period of thirty days if, during the course of this period, they have not been confirmed by the Minister of Public Health or by his delegated official; the individuals affected by the measures in question must previously be consulted or called.

6.1.4.6 SANCTIONS

Penal sanctions provided for under article 11 of the law impose imprisonment of from eight days to six months and a fine of from 26 to 5,000 francs, or one of these punishments only, on the following:

(1) those who, not being authorisation holders, carry out acts or activities which are subject to prior authorisation;
(2) those who carry out acts or activities which are prohibited under the law;

(3)   those who, through negligence or lack of forethought in their use of mobile or immobile objects, are the cause of an alteration to the groundwaters, thereby making them unsuitable for water supply and domestic use;

(4)   those who refuse or oppose inspection visits, the taking of samples or the measures contained in articles 9 and 10 (see 6.1.4.5).

Article 11(2) deals with the repression of repeated offences and article 11(4) allows the judge to order the demolition of installations and works constructed in contravention of the provisions issued in application of the law. He can even order the site to be restored to its original natural state.

If the offender fails to carry out the judgement within the specified period, it will be carried out officially at his own risk, through the intervention of the Minister of Public Health.

Companies are responsible in civil law for all financial and confiscatory judgements pronounced against their representatives or employees for infringements of article 11 of the law. These companies can be summoned to appear before the criminal courts.

6.1.4.7

It should be repeated at this point that at present the Law of 26 March 1971 on the protection of groundwaters is still not observed. Other legislative texts on the subject should therefore be mentioned.

6.2 QUALITATIVE ASPECTS

The quality protection of groundwaters is also the subject of the following legal texts:

6.2.1 The Law of 1 August 1924, concerning the protection of mineral and thermal waters[2]

This law enables the government to declare that it is in the public interest to protect a source of mineral or thermal water belonging to the State, a province, a commune or an association of communes (article 1).

The royal decree which recognises the public interest determines an area

113

in which no work which could result in reduction of the source of water supply, or in alteration of the quality of the water supplied, can be implemented without prior authorisation (article 2).

The area of protection thus identified and the classification of works requiring prior authorisation can be further modified by a new royal decree.

The following articles of the law determine the procedure which should precede the decree envisaged by article 1, as well as the procedures to be followed by the authorisation applications envisaged in article 2, and the sanctions which can be applied to those who contravene this article.

6.2.2 The Health Law of 1 September 1945

This authorises the King to prescribe, after consultation with the Superior Council of Public Hygiene, through general regulations the measures for prophylaxis and cleansing as well as *all* organisational and control measures necessary to ensure the supply of drinking water in particular.[3]

The possibilities offered by this law are (also) little used.[4]

6.2.3 The Law of 20 June 1964, concerning the inspection of foods, food products and other products

As a result of this law, a Royal Decree of 24 April 1965 fixes the quality standards for water intended for human consumption which is distributed through the water supply network.[5]

6.2.4 Rural Code, article 90, clause 2

This clause punishes with a fine of from fifteen francs to twenty-five francs—to be multiplied by sixty—and by imprisonment of from one to seven days, or by one of these punishments alone, those who 'have voluntarily thrown or caused to be thrown into a public or private well, pond or fountain, organic bodies or any other substance of a kind to taint the water or to make it unclean for domestic use'.

6.3 QUANTITATIVE ASPECTS

6.3.1 Introduction

The quantitative protection of groundwaters is the subject of:

(1) the Royal Decree of 18 December 1946 instituting a census of groundwater reserves and establishing controls for their use;[6]
(2) the Royal Decree of 21 April 1976 concerning authorisation of new groundwater offtakes;
(3) the Law of 9 July 1946 relating to the control of the exploitation of groundwaters.[7]

6.3.2 Rights conferred on the King

The Decree of 18 December 1946 confers on the King the right to:

(1) Fix the conditions to which authorisation for establishing any new groundwater offtake or any installation using groundwater will be subject. The expression 'groundwater offtake' applies to all wells, catchment drainage, and in general, all works and installations intended for the abstraction of groundwater, including emergency water catchment.
(2) Control the use of casual sources of groundwater produced during the course of works carried out underground such as, for example, the exploitation of quarries, mines or pits.
(3) Fix the conditions in which a general national census of groundwater resources will be carried out.

6.3.3 The use of groundwaters

The Royal Decree of 21 April 1976[8] makes all *new* groundwater offtakes subject to prior authorisation.

The following are exempt from authorisation provided that the water is not welling up at the place of catchment:

(a) Groundwater offtakes intended for the sanitation and drinking water purposes of a family community, as well as for the necessary space cleaning of the area which it occupies.

(b) Wells where the water is abstracted without the assistance of an engine.

(c) Trial pumping of a duration of not more than two months in order to find out the characteristics of an aquifer or to predetermine the characteristics of a future catchment operation, provided that the water abstracted is not for industrial use and is not introduced into the water supply network.

(d) The pumping of groundwater of a temporary nature carried out during public or private construction of civil engineering works when the output collected does not exceed 96 m³ per day.

An authorisation is also required for any alteration to a groundwater offtake and for any conversion or resumption of use of a groundwater offtake after a continuous period of inactivity lasting for two years or more.

A distinction is made between two types of authorisation. In *Class I* are groundwater offtakes with a daily abstraction rate not exceeding 96 m³, as well as temporary pumpings carried out during public or private construction or civil engineering works with a supply of more than 96 m³ per day.

Class II comprises groundwater offtakes with a daily abstraction rate exceeding 96 m³, with the exception of the temporary uses already mentioned.

Groundwater offtakes in Class I require authorisation from the engineer of mines and can be subject to conditions which must be justified. For groundwater offtakes in Class II, authorisation is granted by the Minister responsible for mines, quarries and pits, and must be authorised by justified decree. At any time, by justified decision, new conditions can be imposed on both classes of groundwater offtake.

There is an obligation to make a declaration to the director general of mines before any tests for the exploitability of an aquifer can be carried out. The director general of mines acknowledges receipt of the declaration and can make the test subject to certain conditions.

6.3.4 Offtakes established before 1947, and those having an abstraction rate in excess of 96 m³ per day

Since the Royal Decree of 21 April 1976 only authorised the definition of conditions for the installation of new water offtakes, on 9 July 1976 a law was promulgated, authorising the King to control the exploitation of groundwater offtakes established before 15 July 1947 which cannot be assimilated to new groundwater offtakes.[9]

A further Royal Decree of 1 October 1976 determines when the minister responsible for mines, quarries and pits can impose exploitation conditions on those groundwater offtakes having an abstraction rate in excess of 96 m^3 per day.

6.3.5 The use of casual sources of groundwaters

The use of such casual sources arising from the exploitation of mines other than coal mines, quarries pits and underground excavations is controlled by a Royal Decree of 13 July 1976.[10]

6.3.6 The census of groundwater resources

The census of groundwater resources was organised by the Royal Decrees of 14 June 1966 and 9 August 1976.[11]

6.3.7 Supervision of the application of the Decree of 18 December 1946

Under the Royal Decree of 21 April 1976[12] the officials of the Mines Administration designated for the purpose by the minister in charge of mines[13] are responsible for the supervision of the application of the Decree of 18 December 1946.

They have access at all times to works and installations under their supervision. They ascertain any infringements of the above mentioned decrees by written report which is used as evidence in the absence of proof to the contrary.

6.3.8 Sanctions

Infringements of the provisions of the Decree of 18 December 1946, the Law of 19 July 1976 and of their implementing decrees are punished with a fine of from 500 to 2,500 Belgian francs.

Company owners, operators, directors, managers or other employees who have obstructed the supervisory activities of the agents of the administration will be punished with a fine of from 10 to 1,000 francs.

In the event of a repeated offence within twelve months from a previous judgement, the minimum and maximum fines can be doubled.

The owners, users and operators of companies are responsible under civil law for the fines inflicted on their directors, managers or other employees.

As a result of article 4 of the Decree of 18 December 1946, the judge can order the confiscation of machines and the demolition of works as well as the restoration of the site to its original natural state, under the supervision and in accordance with the requirements of the Mines Administration. The Law of 9 July 1976 contains no similar requirements.

Notes

1. A royal decree will fix—'being careful to preserve the rights of the defence'—the methods by which samples are to be taken, the rules for the approval procedure as well as the model or protocol for analyses. It can also fix the analysis methods.
2. *Moniteur belge*, 22 August 1924.
3. *Moniteur belge*, 10 October 1945.
4. H. Coremans, Het water also onderwerp van overheidszorg en wetgeving, *Tijdschr. Bestuursw. Publiekr.*, 1970, p. 93.
5. *Moniteur belge*, 16 June 1965; this royal decree was completed with the Royal Decree of 6 May 1966, *Moniteur belge*, 13 July 1966.
6. *Moniteur belge*, 6 March 1947.
7. *Moniteur belge*, 28 August 1976.
8. *Moniteur belge*, 25 June 1976.
9. *Moniteur belge*, 15 October 1976.
10. *Moniteur belge*, 21 August 1976.
11. *Moniteur belge*, 6 July 1966, and 4 September 1976; see also the Ministerial Decree of 29 July 1966, *Moniteur belge*, 21 September 1966, and the Ministerial Decree of 21 November 1973 relating to metering devices for groundwater offtakes.
12. *Moniteur belge*, 25 June 1976.
13. See article 1 of the Ministerial Decree of 11 August 1976 which designates agents at levels 1 and 2 of the Mines Administration, including those of the Belgian Geological Service. These same agents are responsible for the supervision and application of the Law of 9 July (Ministerial Decree of 25 November 1976).

7

The Protection of the Sea

It is obvious that the protection of the sea as a whole against the risks of pollution arises primarily from international environmental law.

There are numerous international conventions which have been signed during the course of recent years.

Although Belgium is generally included among the signatory States to such conventions both individually and as a member of the EEC, it should be remembered that the ratification of these agreements is sometimes an unreasonably long (and) drawn out procedure. It is therefore not surprising that this chapter is a slim one.

Among the most important international conventions awaiting ratification, the following in particular should be noted:

The International Convention for the Prevention of Pollution by Ships, London, 2 November 1973.

The Convention on the Prevention of the Pollution of the Sea by the Dumping of Waste, London, 29 December 1972.

In this chapter we shall first discuss the conventions ratified by Belgium concerning hydrocarbon pollution. We shall subsequently examine the international texts ratified by Belgium concerning the prevention of marine pollution from dumping operations. Finally, the exploration of the continental shelf, the exploitation of natural resources and the protection of coastal waters will be considered.

7.1 HYDROCARBON POLLUTION

7.1.1 The London Convention of 1954

7.1.1.1

Through the Law of 29 March 1957, Belgium approved the international convention of London signed by thirty-one States on 12 May 1954 and in this way became associated with the international plan of action for the prevention of hydrocarbon pollution of the sea. The amendments to the Convention adopted in London on 11 April 1962 and 21 October 1969 were approved in Belgium by the Laws of 14 January 1966 and 19 March 1973 respectively.

This convention instituted a system of zones in which certain ships are forbidden to discharge hydrocarbons. The necessary sanctions to enforce such a prohibition are also provided for.

7.1.1.2

The Law of 4 July 1962 on the hydrocarbon pollution of sea water was adopted on the basis of the London Convention of 1954, as amended in 1962 and 1969. This law was implemented by the Royal Decree of 29 November 1967, modified by the Royal Decree of 12 September 1978.

This internal legislation applies to sea-going ships flying the Belgian flag or belonging to the Belgian State with the exception of ships belonging to the Navy or used as auxiliary ships by the Navy. Ships which do not use fuel-oil or heavy diesel oil for their propulsion and do not transport hydrocarbons are also excluded.

7.1.1.3

The Law of 4 July 1962 contains a general prohibition on discharge of hydrocarbons or any mixture containing hydrocarbons liable to dirty the surface of the sea within the limits of the zones defined in the convention, its annexes, or yet to be determined as a result of the convention.

All ships other than tankers to which the London Convention applies are forbidden to discharge hydrocarbons or hydrocarbon mixtures, except on the following conditions:

(a)   the ship is travelling;
(b)   the instantaneous discharge rate of hydrocarbons does not at any time exceed 60 litres per nautical mile;
(c)   the hydrocarbon content of the discharge is less than 100 ppm of the mixture;
(d)   the discharge is carried out as far as possible from land.

All tankers[1] to which the Convention applies are forbidden to discharge hydrocarbons or hydrocarbon mixtures except on the following conditions:

(a)   the tanker is travelling;
(b)   the instantaneous discharge rate of hydrocarbons does not at any time exceed 60 litres per nautical mile;
(c)   the total quantity of hydrocarbons discharged during the course of a ballast voyage does not exceed 1/15,000 of the total capacity of the cargo holds;
(d)   the tanker is more than 50 nautical miles from the nearest land.

This prohibition does not apply to the discharge of ballast waters from a cargo tanker which has been cleaned since the delivery of its last cargo provided that its effluent leaves no apparent trace of hydrocarbons on the surface of the water when it is discharged into calm waters in fine weather.

7.1.1.4

There are two important exceptions to this discharge prohibition which concern both tankers and other ships:

(a)   hydrocarbons or hydrocarbon mixtures discharged by a ship for the purpose of ensuring the safety of the ship, to prevent damage to the ship or its cargo, or to safeguard human life at sea;
(b)   the loss of hydrocarbons or hydrocarbon mixtures following damage to the ship or an unavoidable loss if all reasonable precautionary measures have been taken to avoid such loss or reduce it to a minimum following the discovery of the damage or loss.

7.1.1.5

The prohibitions described under 7.1.1.3 apply to all ships, regardless of their flag, if they are in Belgian territorial or maritime waters (see 7.4.1).

In other words, the system is as follows: the regulation applies on the high seas to ships flying the Belgian flag and to all ships within the limits of Belgian territorial waters since Belgium is the territorial State.

7.1.1.6

The Law of 4 July 1962 enables the King to:

determine the manner in which ships should be equipped in order to avoid the flow of hydrocarbons towards the outlets when the contents are to be discharged into the sea without pretreatment by an approved oil separator;

make the installation of effective oil separators compulsory on board certain categories of ship.

The Royal Decree of 29 November 1967 contains detailed provisions concerning guttering, drainage tanks, oil greasing tanks, rinsing basins and oil separators.

When ships do not conform to these provisions they can be detained by the maritime inspector and the maritime commissioner.

7.1.1.7

The King appoints ports which must be equipped with collection installations. He also determines the minimum volume of residue that they must be able to receive during a specified period of time.

The appointed port authorities are obliged to ensure that their ports are equipped with these installations. The Cleaning Station in Antwerp has been functioning since 1971.

7.1.1.8

The captain of every ship must keep a hydrocarbon register on board.[2] Within 24 hours of arriving in port, he must inform the maritime commissioner or consul of all discharges or leakages into the sea, whatever the cause, of mixtures containing hydrocarbons likely to dirty the surface of the sea.

The captain must allow the responsible authorities to check the hydrocarbon register on board his ship when it is in port in a territory of which the government is party to the Convention.

He is also required to supply a copy of any portion of the register, on request, and to certify that the copy is accurate. A copy taken under these conditions is taken as fact in the absence of proof to the contrary.

Any intervention by the responsible authorities must be carried out as rapidly as possible and without causing delay to the ship.

7.1.1.9 MONITORING THE IMPLEMENTATION OF THE LAW

Article 8, section 1 of the Law of 4 July 1962 lists the following officials as responsible for investigating and verifying infringements:

(1) maritime commissioners and agents of the maritime police and the maritime gendarmerie;
(2) harbour masters;
(3) officials and agents of the maritime inspection service;
(4) officials and agents of the bridges and roads administration responsible for monitoring shipping routes;
(5) consuls.

In implementation of the London Convention of 1954, the first three categories appointed are authorised to exercise control in the ports of the Kingdom and on board ships of the merchant navy which are subject to the Convention and navigating under the flag of a country which is party to the Convention.

In addition, the officials and agents appointed by other countries party to the Convention are also competent to verify infringements of the provisions of the Convention.

7.1.1.10 SANCTIONS

Infringements of the Law of 4 July 1962 are subject to fines but not personal imprisonment.

Captains and officers who contravene the requirements of article 2 (discharge of hydrocarbons) are punished with a fine of from 1,500 to 7,500 francs. The fine is from 2,500 and 12,000 francs if the infringement is committed between sunset and sunrise.

Captains and officers are liable to a fine of from 100 to 250 francs if they contravene the requirements of the law concerning the hydrocarbon register (article 10).

A captain who contravenes the obligation to allow the authorities to consult the hydrocarbon register when they request to do so incurs a fine of from 250 to 500 francs.[11]

A captain who fails to make the necessary declaration to the maritime commissioner or consul is punished by a fine of from 100 to 250 francs (article 12).

Finally, a captain and owner of a ship who infringe the requirements relating to equipment to prevent the flow of hydrocarbons towards the outlets or the installation of effective oil separators (article 13) are fined respectively from 250 to 1,000 francs or from 1,000 to 5,000 francs.

7.1.1.11 COURTS

The responsible court is that of the area in which the ship's home port is situated.

Any Belgian or foreigner on board a Belgian ship who infringes the provisions of the law in question or its implementing decrees outside Belgian territorial limits is liable to prosecution in Belgium.

7.1.2 Civil liability: The Brussels Convention of 1969

7.1.2.1

The Law of 20 July 1976 approves and implements the international convention and its annex, concluded in Brussels on 29 November 1969[3] on civil liability for damage caused by hydrocarbon pollution. This law fixes the conditions to be fulfilled with regard to insurance and financial guarantees. It establishes a series of penal provisions (fines). The law states that the provisions of article VII of the Convention apply to all ships registered in Belgium, in addition to all ships entering or leaving a Belgian port.

All Belgian ships transporting more than 2,000 tonnes of hydrocarbons in bulk as cargo by sea, including territorial waters, must carry a certificate of financial guarantee granted by the Minister of Communications, on the favourable opinion of the Minister of Economic Affairs. A Royal Decree of 10 March 1977 and a Ministerial Decree of 28 March 1977 specify the form and content of the certificate.

The court of first instance in Brussels is the only court which can judge compensation questions for damage caused by pollution on national territory, including territorial waters, as a result of article IX of the Convention.

7.1.2.2

The following are responsible for investigating and verifying infringements:

(1) maritime commissioners and agents of the maritime police;
(2) harbour masters;
(3) officials of the customs and exercise administration;
(4) Belgian consular officials with regard to ships registered in Belgium.

When information is lodged concerning a serious infringement the maritime commissioner can, if necessary, in view of the copy of the report sent to him by the informing authorities, arrest the ship transporting hydrocarbons and one or several other ships belonging to or operated by the same person, for as long as necessary, at the cost and risk of the owner or operator.

This measure can only be lifted by the maritime commissioner when all the obligations arising from the Convention and national legislation have been met. In addition, proof must be provided that a bail payment equal to the most severe fine envisaged under article 7—to be multiplied by sixty—has been paid to the cashier ('Caisse des Dépôts et Consignations'). The interest payable on the sum deposited is added to the bail payment.

7.1.3 Accidents which can involve pollution of the sea by hydrocarbons

7.1.3.1

With regard to accidents involving, or which could involve, the pollution of the sea by hydrocarbons, the Law of 29 July 1971[4] approved the Convention of 29 November 1969 on Intervention at Sea in the Event of Accidents.

In addition, Belgium is party to the agreement of 9 June 1969 regarding co-operation concerning the fight against the pollution of the North Sea;[5] this agreement was completed by a technical arrangement on 28 July 1972 between Belgium, France and Britain,[6] which organises the monitoring of the zone between 51°32 and 51°06 latitude North.

The chief objective of this agreement was to organise the exchange of information between the party States. The North Sea is divided into zones; the contracting party for the zone in which an accident occurs or in which an oil slick is spotted must make the necessary assessment concerning the nature and extent of the accident or, if necessary, the type and approximate quantity of hydrocarbon floating on the sea, as well as the direction and speed of movement of the slick, and must immediately inform all the other contracting parties.

7.2 PREVENTION OF MARINE POLLUTION FROM DUMPING OPERATIONS

7.2.1

The Law of 8 February 1978 approves the Convention for the Prevention of Marine Pollution from Dumping Operations Carried Out by Ships and Aircraft, and its annexes, concluded in Oslo on 15 February 1972.[7]

The Oslo Convention is a regional convention. It applies to the north-east area of the Atlantic and part of the Arctic Ocean.[8]

The preventive character of the convention is made clear in its first article, which requires the contracting parties to promise to take all possible measures to fight against sea pollution by substances liable to endanger the health of man, to damage biological resources or marine fauna and flora, to harm amenities or obstruct any other legitimate uses of the sea.

The convention is based on a system of a 'black' list (prohibition to dump) and a 'grey' list (authorisation procedure; specific permit; general permission).

It is forbidden to dump the substances listed in annex I (the black list) of the Convention, except in a critical situation. As a result of article 3 of the Law of 8 February 1978, the King can extend the dumping prohibition to cover other substances in addition to those listed in annex I. It is forbidden to dump waste containing the substances and materials listed in annex II (the grey list) of the Convention without a specific permit which must be granted in each case.

Finally, materials not included in the first two annexes of the Convention cannot be dumped without the approval of the relevent national authority.

Annex III of the Convention sets out the factors to be taken into consideration in granting a permit or approval to dump at sea.

7.2.2

A commission composed of representatives from each contracting party is responsible for:

(a) carrying out general monitoring of the implementation of the Convention;

(b) receiving and evaluating the lists of permits and approvals granted for dumping operations and defining the model procedure to be followed for this purpose;

(c) examining in a general manner the condition of the sea water within the limits of the zone to which the Convention applies; assessing the effectiveness of the control measures adopted and the need for any additional or different measures;

(d) up-dating the annexes of the Convention and recommending modifications, additions or deletions which should be adopted;

(e) fulfilling any other functions required under the terms of the Convention.

7.2.3

The Law of 8 February 1978 established a series of penal provisions. Fines can be applied of up to one million Belgian francs.

The Law of 8 February 1978, as does the Law of 20 July 1976 already mentioned, empowers the maritime commissioner and, in the context of this law, the inspector of aeronautical police, to arrest the ship or aircraft in question in the event of infringement. This measure cannot be lifted until all the obligations arising from the Convention and national legislation have been fulfilled and proof has been provided that a bail payment equal to the most severe fine has been paid to the cashier. The interest payable on the sum deposited is added to the bail payment.

When the responsible authorities have proof that there are substances or materials on board a ship or aircraft that are to be dumped at sea in infringement of the provisions of the Convention, the law or its implementing measures, or when they have serious reason to believe that this is the case, the owner or operator of the ship or aircraft, their agents or representatives are required, if requested to do so, to head immediately for a port or airport designated by the authorities.

If this request is not complied with, the relevent authority can take the necessary measures to arrest the ship and have it brought into the appointed port at the cost and risk of the owner or operator, their agents or representatives. The ship can be pursued on the open sea, if necessary, and brought back under the same conditions.

7.3 THE CONTINENTAL SHELF

7.3.1

Article 24 of the Geneva Convention of 29 April 1958 on the open sea invites States to adopt adequate controls to avoid sea pollution resulting from the exploration and exploitation of the sea bed and sub-bed:

> Every State is required to issue rules aimed at avoiding sea pollution by hydrocarbons dispersed by ships or pipelines or resulting from the *exploration or exploitation of the sea bed or sub-bed,* while taking account of the conventional provisions which already exist on the subject.

On the same day, 29 April 1958, a convention was signed in Geneva concerning the continental shelf. Belgium has not signed this convention,[9] but, by means of an internal law has confirmed its national rights to the continental shelf in the North Sea. This law was passed on 13 June 1969 and adopts those provisions of the Geneva Convention which are considered the most appropriate for the Belgian continental shelf.

7.3.2

Under the Law of 13 June 1969:

(a) the term 'continental shelf' means the sea bed and sub-bed of submarine areas adjacent to the coast but outside territorial waters;
(b) 'natural resources' include mineral resources and other non-living resources of the sea bed and sub-bed, as well as living organism belonging to sedentary species, in other words, organisms which, when caught, are either stationary on or above the sea bed, or incapable of moving if they are not in constant contact with the sea bed or sub-bed.

The research and exploitation of mineral resources and other non-living resources of the sea bed and sub-bed are subject to the granting of concessions in accordance with the conditions and procedures determined by the King.

128

Installations and other devices situated offshore on the continental shelf and covered by the law, as well as people and property located on such installations or devices, are subject to Belgian law.

7.3.3

The King must ensure that exploitation and exploration do not constitute obstacles to navigation, fishing, the conservation of the biological resources of the sea or scientific research. These provisions are in accordance with those contained in article 5 of the Geneva Convention of 29 April 1958.

The King determines all requirements he considers necessary for this purpose, particularly with regard to signalling, the means of avoiding pollution and damage to submarine cables or pipelines.

7.3.4

The Royal Decree of 16 May 1977 was issued as a result of the Law of 13 June 1969. It contains provisions for the protection of navigation, maritime fishing, the environment and other essential interests during the exploration and exploitation of mineral resources and other non-living resources on the sea bed and sub-bed in territorial waters and on the continental shelf.[11] Arricle 2, s.3, states:

> During the course of exploration, exploitation and associated activities, all measures must be taken to avoid all forms of pollution. In the event of accident, the necessary measures to reduce damaging consequences must be taken immediately.

A 500 metre safety zone is marked out around each offshore anchored or fixed installation or device.

The safety conditions can be modified during the course of exploration by royal decree on the proposal of the Minister of the Mines Administration, after consultation with the concessionaire and with the opinion of the ministers involved in granting concessions.

In the event of imminent danger or if the holder of the concession or authorisation refuses to comply with the legal, regulatory or concessionary conditions, the ministers responsible, or their appointed officials, can take the necessary measures to safeguard the safety of shipping, maritime fishing, the environment and other essential interests.

These measures must be implemented by the holder of the concession or authorisation at his own cost and risk, within the period of time specified by the ministers or their appointed officials.

If these measures are not observed, or the time for their implementation is exceeded, the Minister of the Mines Administration can proceed with the total or partial suspension or cancellation of the concession.

If an installation or device used for offshore exploration or exploitation ceases operations permanently, it must be removed at the cost and risk of the concessionaire.

7.4 THE POLLUTION OF TERRITORIAL WATERS

7.4.1

It should be remembered that the Law of 26 March 1971 on the protection of surface waters against pollution[12] also covers the protection of coastal and territorial waters against pollution.

Coastal waters are taken to mean territorial waters; in other words, the coastal seas up to a distance of three nautical miles.

7.4.2

The Royal Decree of 31 May 1968 acting as a police and navigation regulation for Belgian territorial waters, coastal ports and beaches, contains the following provision regarding the prevention of pollution:

'Article 13, s.2: it is forbidden:

(1) to throw, deposit, or allow any object which could raise the sea bed, hinder navigation or interfere with the free flow of water to float or flow into Belgian territorial waters, coastal ports or beaches; to *discharge any matter or liquid* and particularly hydrocarbons, hydrocarbon residues or radioactive waste *likely to pollute the waters;* captains or owners must conform to the orders on the subject issued by the officials or agents of the authority'.

Notes

1. 'Tankers' are taken to mean all ships in which the greatest space is used for cargo, constructed or adapted for the purpose of transporting liquid in bulk and which, at the moment in question, only transport hydrocarbons in the space reserved for cargo (article 1, Royal Decree of 29 September 1967).
2. The model was determined by the Ministerial Decree of 30 November 1967 and amended by Ministerial Decree of 18 October 1974.
3. *Moniteur belge*, 13 April 1977; err. *Moniteur belge*, 26 June 1979. The Convention applies to damage incurred on the territory of the contracting States; 'damage' includes material damage directly caused by the pollution and the cost of measures taken to avoid or reduce the damage.
4. *Moniteur belge*, 2 February 1972; entry into force: 6 May 1975. *Moniteur belge*, 30 April 1975.
5. *Moniteur belge*, 23 October 1969.
6. *Moniteur belge*, 13 March 1973.
7. *Moniteur belge*, 4 May 1978.
8. 'Since the Convention extends the exclusive rights of the coastal State to living organisms in permanent contact with the sea bed (oysters, mussels, algae etc.), the absence of any concrete limiting criteria could cause serious injury to our deep sea fishermen. This has led the Belgian government to abstain from signing the convention of 29 April 1958 on the continental shelf' (Doc. Parl. Chambre des Représentants, 1966–1967, no. 471–1, exposé of reasons).
9. Norway acted similarly with the Law of 21 June 1963.
10. *Moniteur belge*, 21 July 1977.
11. See Chapter 5.
12. The ports of Ostend, Zeebrugge, Nieuport and Blankenberge.

8

The Treatment and Disposal of Waste

8.1 INTRODUCTION

The consumer society as we know it today has to deal with a super-abundance of residual waste products of which a large number can be degraded only with difficulty or not at all.

It is essential to set up a comprehensive management policy based on prevention, recovery and recycling of raw and secondary materials and the disposal of wastes in order to deal with the growing quantity of all types of waste.

Belgium has no outline law on the subject and there is a wide variety of spheres of responsibility and regulatory provisions. The Council Directive (75/442/EEC) of 15 July 1975[1] fixing the general objectives and proposing a series of specific provisions should have been introduced into the legal texts in July 1977, but as a result of the problems relating to devolution of powers to the Regions, Belgium has been found in contravention of community rules on the subject by the Court of Justice of the European Communities, in a judgement of 2 February 1982, case 69/81.

Actually, national parliament no longer has the power to adopt general legislation on waste; the regional councils are now responsible (see Chapter 1). 'Even with regard to the management of toxic wastes and radioactive wastes, the principle of regional responsibility applies, with the provision that certain aspects of this management relate to matters reserved for the national state, i.e. the general and sectoral legal provisions concerning the protection of the environment and the nuclear fuel cycle.'[2]

8.2 TOXIC WASTES

Following the discovery of clandestine deposits of toxic waste, which were subsequently taken over by the authorities, a strict and rigorous legal instrument was introduced in Belgium. The Law of 22 July 1974, completed by the Royal Decree of 9 February 1976, lays down the principal provisions for control.[3]

This law is aimed at protecting the population, workers and the environment in general from the dangers of toxic waste.

In the first place, it gives a definition of 'toxic waste'. It then provides for a whole series of prohibitive clauses and obligations to be imposed on the producers of wastes and for a system of authorisation. Finally, it formulates the sanctions intended to ensure that the obligations imposed will be respected. Council Directive 78/319/EEC of 20 March 1978[4] was largely inspired by the Belgian legislation.

8.2.1 Definition

The first article determines what is meant by toxic wastes: these are unused or unusable products or by-products, residues and wastes resulting from an industrial, commercial, craft, agricultural or scientific activity which could present a danger of intoxication for living beings or nature.

Article 2 of the Royal Decree of 9 February 1976 acting as a General Regulation on toxic wastes contains a list of substances which are considered toxic.

It should be noted that this list includes packaging which has contained toxic wastes or has been polluted by the wastes but is no longer used.

Any modifications to this list must be submitted for an opinion to the Commission of Approval, the membership of which is laid down in article 8.[5]

8.2.2 Means of prevention and control

8.2.2.1 PROHIBITIVE CLAUSES

The dumping of toxic wastes is absolutely forbidden. 'Dumping' is taken to mean the dumping of wastes outside the place where they are made or used, in any place whatsoever.[6]

The following are also forbidden except by authorisation or declaration: the offer for sale or sale, the purchase, the free or conditional gift, the holding, storing, processing, destruction, neutralisation or disposal of toxic wastes, as well as all other affiliated activities.

The violation of this prohibition is punishable by imprisonment of from eight days to one year and by a fine of from 100 francs to 100,000 francs, or by one of these punishments alone. In the case of a repeated offence, the punishment can be increased to double the maximum.

8.2.2.2 AUTHORISATION AND DECLARATION REQUIREMENTS

The methods of procedure concerning the application and granting of authorisation, as well as the general conditions for the approval of such authorisation, have been fixed by the Royal Decree of 9 February 1976. The authorities are also designated in the decree.

8.2.2.2.1 Operating authorisation

The destruction, neutralisation or disposal of toxic wastes can take place either in installations belonging to the producer of the toxic wastes or in centres approved by the King, called Centres for the Destruction, Neutralisation or Disposal of Toxic Wastes. The application for approval is introduced at the same time as the application for an operating authorisation.

The authorities granting the operating authorisation, and the procedures to be followed, are the same as those laid down for Class I establishments in Chaper 1, Heading I, of the General Regulation for Protection at Work.[7]

In addition to the information required under the General Regulation for Protection at Work, the authorisation application must contain information on the nature and methods of disposal envisaged for residues from the treatment of toxic wastes.

For decisions concerning toxic waste centres, the authority granting the authorisation must previously obtain the opinion of the Commission of Approval. In the event of an unfavourable opinion, the operating authorisation is refused.

The Commission is required to decide within two months from receipt of the application.

8.2.2.2.2 Authorisation for the purchase and importation of toxic wastes

The purchase and importation of toxic wastes can only be carried out by:

approved centres for the destruction, neutralisation and disposal of toxic wastes;

persons who import or purchase toxic wastes for the purpose of destruction, neutralisation or disposal in approved centres, who are authorised to purchase and import such substances;

persons who purchase toxic wastes for the purposes of exporting them, who have a purchase authorisation.

The authorisation application is sent to the Administration for Hygiene and Medicine at Work; it is subject to the opinion of the Commission of Approval, if a professional activity is involved. The Commission can put forward conditions to which the authorisation should be subject.

When a professional activity is involved, the authorisation is granted by the Minister of Employment and Labour for a maximum period of ten years and is renewable.

For applications concerning a single purchasing or importation operation, authorisation is granted by the Administration for Hygiene and Medicine at Work.

When an authorisation holder fails to observe the requirements of the General Regulation on toxic wastes or the conditions imposed in the authorisation, the Minister for Employment and Work can suspend or withdraw that authorisation.

8.2.2.2.3 Royalties levied by the State

A royalty is levied by the State at the time of presentation of an application for authorisation for the purchase or importation of toxic wastes, as follows:

— 10,000 francs for a professional activity;

— 1,000 francs for a single operation.

8.2.2.2.4 Declaration of detention, sale, conditional or free gift, purchase, importation or exportation of toxic wastes

There is an obligation to make a declaration to the Administration of Hygiene and Medicine at Work:

within eight days from the commencement of operations in the case of detention, sale, conditional or free gift, or exportation; declarations can be made periodically for professional activities;

monthly in the case of purchase or importation.

8.2.2.2.5 Sanctions and waivers

The non-observance of the authorisation or declaration obligations, or conditions imposed is punishable by imprisonment of from eight days to one year and a fine of from 100 to 100,000 francs, or by one of these punishments alone. The punishment can be increased to double the maximum in the case of a repeated offence within three years.

In exceptional circumstances the ministers responsible can grant waivers of the Royal Decree of 9 February 1976 by means of decrees giving reasons for the waiver.

8.2.2.3 THE DESTRUCTION, NEUTRALISATION OR DISPOSAL OF TOXIC WASTES

8.2.2.3.1 Obligations concerning the producers of toxic wastes

Individuals exercising an industrial, commercial, agricultural, craft or scientific activity which produces toxic wastes are obliged to destroy, neutralise or dispose of these wastes at their own cost.

The law upholds the principle that the person carrying out an activity which produces toxic wastes is, and remains, wholly responsible for these wastes, even if he has conferred the transport, destruction, neutralisation or disposal on another person.[8]

In consequence, the producer of toxic wastes is always responsible, even when he himself is not at fault. For this reason an injured party does not require proof that a fault has been committed in order to obtain compensation.

This is an example of strict liability; see Chapter 12.

8.2.2.3.2 Installations

The destruction, neutralisation or disposal of toxic wastes should be carried out either in installations belonging to the producer of the toxic wastes, or in installations in a 'centre' approved by the King, on the proposal of the Minister for Employment and Labour.[9]

The operator of a business where these activities take place is obliged to take the necessary steps to ensure that they do not become a source of danger or inconvenience for personnel, neighbours, the public or the environment in general. There are also safety provisions regarding the packaging and transport of toxic wastes.

The King can, at any time, suspend or withdraw approval if a centre for the destruction, neutralisation or disposal of toxic wastes does not observe the provisions of the General Regulation on toxic wastes and the conditions imposed in the approval decree or in the operating authorisation decree. Additional conditions can also be imposed.

For all these decisions the prior opinion of the Commission of Approval is required.

8.2.2.3.3 Packaging

The law also contains a provision authorising the government to require packaging containing poisonous, soporific, narcotic, disinfectant or antiseptic substances to include indications on the methods for their destruction, neutralisation and disposal.[10]

This clause is extremely important; it enables the consumers of these products to be aware of the methods of getting rid of their wastes in such a way that they do not constitute a danger to health or to the environment.

8.2.2.3.4 Sanctions

The non-observance of the prescriptions on the subject of the destruction, neutralisation or disposal of toxic wastes is punishable with imprisonment of from eight days to one year and a fine of from 100 to 100,000 francs, or by one of these punishments alone.

8.2.3 Methods of supervision and intervention

8.2.3.1 SUPERVISION

8.2.3.1.1 Qualification of personnel in charge of the operation

The employer is obliged to designate a person who will be responsible for the destruction, disposal or neutralisation of toxic wastes. The Royal Decree of 9 February 1976 defines the qualification requirements to which this person must conform in order to obtain approval from the minister in charge.[11]

This person is responsible for supervising the observance of the regulations and authorisation conditions under the present law. He controls and supervises the implementation of all the measures which are recognised to be necessary to ensure the safety of workers, the works and the environment.

He enjoys special protection with regard to his employer in the event of dismissal.[12]

8.2.3.1.2 Powers of agents and officials designated by the King

Without detracting from the powers of the judiciary police officers, officials and agents designated by the King supervise the implementation of the law and its decrees.[13]

These officials have free access at all times to all places in which toxic wastes are found. For residential premises, however, authorisation from the judge of the police tribunal is required. Places in which the presence of toxic wastes can reasonably be suspected are equally accessible in conforming to the limits indicated in the law.

They can interrogate individuals, examine documents, take photocopies of them and even remove them. They can also take samples in order to determine the composition of the waste.

The method of sampling and the procedures to be followed for the analyses have been fixed by the Royal Decree of 9 February 1976; it also provides for approval of several analytical laboratories.

They are also given the following responsibilities:

to issue warnings;

to fix the period during which an offender can regularise his position;

in the event of infringement, to seal off or seize the toxic wastes, even if the holder of the wastes is not their owner, as well as any means of transport which have been used in committing an offence;

in the event of infringement, to draw up a written report. One copy of this report must, if it is to be valid, be sent to the offender within fourteen days from the verification of the infringement.

These officials and agents can, in carrying out their duties, request the assistance of the communal police and the gendarmerie.

8.2.3.2 METHODS OF INTERVENTION

When it is ascertained that toxic wastes have been dumped, *or* when toxic wastes are the subject of activities for which an authorisation or declaration is compulsory but has not been obtained, *or* when the conditions imposed have not been observed, *or* when the declaration has not beem made, *or* when the toxic wastes have been transported, exported or carried in violation of the regulations issued by the King, the governor of the province where the infringement has taken place can impose conditions, or have the toxic wastes seized, destroyed, neutralised or disposed of.

He can requisition the necessary installations for the disposal of these toxic wastes. The resulting expenditure is borne entirely by the person responsible for the wastes.

If the presence of toxic wastes runs the risk of constituting a serious danger within the commune, independently of any infringement, the governor of the province and the burgomaster of the commune involved can order the transfer of such wastes to a location designated by them or by the Minister of Employment and Work. To this end they can proceed with any necessary requisitions and can request the assistance of the armed forces, the gendarmerie or civil defence.

In the event of an accident or the presence of imminent danger of an accident due to toxic wastes, the Minister of Employment and Labour, in agreement with the Minister of Public Health, the Minister of the Interior, the governor of the province or the burgomaster, decides on the measures to be taken to guarantee the safety of the population or the protection of the environment.[14]

8.2.3.3 SANCTIONS

Individuals obstructing supervision measures or infringing measures taken by the governor of the province or the burgomaster are punished by imprisonment of from eight days to one year and by a fine of from 100 to 100,000 francs, or by one of these punishments only. In the event of a repeated infringement within three years, one punishment can be increased to double the maximum.

8.2.4 Guarantee fund for the destruction of toxic wastes

The law provides for the creation of a fund which is a public company in which the State is authorised to participate.

This guarantee fund has a dual function:

a commercial and industrial objective; that is, to encourage the creation, reorganisation or extension of enterprises for the destruction, neutralisation or disposal of toxic wastes, with registered offices and operational establishments in Belgium, and to ensure the management or possibly acquire capital participation in these enterprises;

an intervention function guaranteeing that in the event of bankruptcy of persons responsible for the destruction of toxic wastes, their obligations can nevertheless be financed by the fund.

This fund is an interesting development because it enables the State to pursue an active policy of encouragement or takeover when private enterprise fails.

8.2.5 Sanctions

We have previously mentioned the punishments applicable to the non-observation of the legal provisions and implementing decrees. In addition, the tribunal can order the temporary or permanent closure of the enterprise. This measure leaves the rights of third parties intact and in particular the rights of the workforce.

The toxic wastes and methods of transport which were used to commit an infringement can be confiscated, even if they do not belong to the person who committed the infringement.

The employer is responsible under civil law for the fines imposed on his employees or representatives.

8.3 NATIONAL LEGISLATION: DOMESTIC WASTE

The collection and removal of household refuse as well as its disposal and the organisation of depots constitutes a responsibility which is traditionally assumed by the communes.

The juridicial basis for the responsibility and obligations of the communes in this field is provided by two revolutionary decrees[15] already mentioned earlier in this report:

Article 50 of the Decree of 14 December 1789: 'The proper functions of the municipal power are . . . to enable the inhabitants to enjoy the benefits of good policing, in particular of *cleanliness,* safety, peace and quiet in the streets, public spaces and buildings'.

Article 3, heading XI of the Decree of 16–24 August 1790: 'The subjects to be dealt with by the vigilance and authority of the municipal services are: everything relating to the safety and comfort of the streets, quays, public places and highways, including the cleaning, lighting, removal of encumbrances, the demolition or repair of buildings in danger of crumbling, the prohibition of placing anything near windows or other parts of buildings which could result in damage from its fall, and the prohibition to throw anything which could damage or injure passers-by or cause harmful fumes'.

These provisions not only confer power on the communal authorities but at the same time they place them under the obligation to ensure that public places and highways are kept clean.[16]

The communes are therefore vested with a general responsibility which includes the power of policing and direct intervention in this area.[17]

'In order to carry out these tasks, the communes can choose one of three management methods:

(1) the constitution of communal services with appropriate qualified personnel and technical equipment;
(2) the creation of an association of communes which can act on behalf of the communes in carrying out the tasks which are allotted to it;
(3) the drawing up of service contracts with specialist companies in the private sector.'[18]

The Royal Decree of 21 December 1960,[19] which specifies in article 1 that the removal of refuse must be ensured by the communal services, whatever the circumstances, enables the burgomaster, district commissioners or governor of the province to requisition for this purpose 'the services of anyone required for the normal operation of the services in charge of the removal of refuse, as well as any essential materials'.

8.4 NATIONAL LEGISLATION: ORGANIC WASTE

8.4.1

The Royal Decree of 24 January 1969 makes the Minister of Agriculture responsible for controlling the collection, disposal and use of organic and kitchen wastes, as well as the sites for the depositing of sludge and dirt.[20]

This decree applies:

to organic waste of animal or vegetable origin;

to kitchen waste (of animal or vegetable origin) arising from collective kitchens or camp sites;

to dirt, that is, all matter containing sludge, organic waste or kitchen waste.

8.4.2

In order to prevent the spread of pathogens to man and animals, as a result of international traffic, the Decree of 24 January 1969 provides for organic and kitchen wastes arising from international transport to be transported either to a disposal plant approved by the Minister of Agriculture, or to an incineration plant approved by the veterinary inspector.

It is forbidden to use these wastes for animal feed, even if they are treated or processed.

Nevertheless, other wastes can be used for animal feed if they have been treated at an approved processing plant or by personnel authorised by the Minister of Agriculture on the proposal of the veterinary inspectors.

8.4.3

The authorisation to establish sites where sludge and refuse can be deposited, refuse depots or composting installations is subject to the prior agreement of the State veterinary inspector.

Sites where sludge and refuse are deposited must be fenced off and access forbidden to unauthorised persons; access is also forbidden to domestic animals. These sites are subject to regular monitoring by the State veterinary inspectors.

The Minister of Agriculture can fix conditions and methods to be used for the cleansing of sites where sludge and refuse are deposited (article 10).

The depositing of organic or kitchen wastes, sludge or refuse on grazing land or land situated along a water course or canal is controlled by the Royal Decree of 17 April 1969.[21]

8.4.4 Sanctions

Infringements of the requirements of the Royal Decree of 24 January 1969 are investigated and punished by sanctions contained in articles 4, 6 and 7 of the Law of 30 December 1882 on the health policing of domestic animals and insanitary insects.

8.4.5

The following should be remembered:

(1) Under the Royal Decree of 11 September 1970,[22] refuse depots and all solid and liquid waste treatment plants have been classified as Class 1 dangerous, dirty or noxious establishments and are placed under the supervision of the Minister of Public Health.

An authorisation is therefore required from the deputed states to erect, convert, relocate or extend a refuse depot or treatment plant for solid or liquid wastes.[23]

(2) Article 6 of the Royal Decree of 26 July 1971, amended by the Royal Decree of 3 July 1972, prohibits the destruction of wastes by burning in the open air in specially protected zones, with the exception of certain wastes of vegetable origin arising from the maintenance of cultivated ground.[24]

8.4.6

Finally, it should be pointed out that the State grants subsidies of up to 60% of the total expenditure to communes which proceed with the construction of installations for the treatment of urban waste.

8.5 NATIONAL LEGISLATION: MISCELLANEOUS

8.5.1 Used cars

Prior to establishing a used car or scrap-iron dump, a 'building permit' is required from the College of Burgomaster and Aldermen.[25] These dumps must be enclosed by walls and trees and must not be visible from the public highway.[26]

With regard to vehicles abandoned on the public highway, the Ministerial Circular of 21 August 1973[27] states that the communes can levy a tax on all such vehicles; when the amount of tax due exceeds the value of the abandoned vehicle, the commune may dispose of it as it sees fit.[28]

8.5.2 Non-toxic industrial wastes

These are generally the responsibility of the Minister of Labour and are contained in List A of the General Regulation on Protection at Work, under the Royal Decree of 18 July 1973, with the exception of certain wastes such as materials arising from building demolition or slag and clinker arising from the metallurgical industry.

Note should also be taken of the Law of 15 September 1979 co-ordinating the laws on quarries, peat bogs and underground mines which takes up certain implementing decrees and provisions relating to mining and quarrying wastes.

8.5.3 Radioactive wastes

See Chapter 9.

8.5.4 The Penal Code

In addition to articles 552(1) and 557(4) of the Penal Code (see 3.4.1.1), article 551(4) of the Penal Code should be noted; this article punishes with a fine of from one franc (!) to ten francs—multiplied by sixty— those who, without necessity or permission from the controlling authority, obstruct the streets, squares or other parts of the public highways, either by leaving materials, scaffolding or any other objects, or by excavations.[29]

8.5.5 The Belgian Waste Pool

In 1978, the Belgian Economic and Agricultural Service—L'Office Belge de l'Economie et de l'Agriculture—was entrusted with the organisation of a Belgian Waste Pool, based on a development of existing pools.

By a 'waste pool' is meant 'an exchange for products which, although classed as pollutants in a specific industry, may be used as secondary raw materials or by-products in another industrial process'.[30]

Initially, the pool serves purely as an information service.

8.6 LEGISLATION FOR THE FLEMISH REGION

The Decree of 2 July 1981 issued by the Flemish Council is an outline decree aimed at stimulating the re-use, recovery and recycling of waste and at controlling the disposal of residual waste.

This decree applies to all waste except:

(a)   radioactive waste (see Chapter 9);
(b)   corpses;
(c)   waste water covered by the Law of 26 March 1971 on the protection of surface waters against pollution (see Chapter 5);
(d)   toxic waste covered by the Law of 22 July 1974 on toxic waste (see 8.1 and 8.2);
(e)   gaseous discharges emitted into the atmosphere referred to by the Law of 28 December 1964 concerning the fight against atmospheric pollution (see Chapter 3).

8.6.1 Organisational framework

8.6.1.1 THE 'PUBLIC COMPANY FOR WASTE IN THE FLEMISH REGION'

The decree sets up a public company called the Public Company for Waste in the Flemish Region. This waste company is authorised by the minister responsible for the removal and treatment of waste in the Flemish Region and it is represented and administered by the Minister; the company is located in Malines and it began operating on 1 October 1981.[32]

The waste company's main task is to work out a co-ordinated waste policy for the Flemish Region. It is also responsible for the planning, control and exploitation of waste.

In addition, the company is required to create a waste data bank.

8.6.1.2 THE WASTE DISPOSAL PLAN

The Flemish Executive works out an overall plan comprising:

(a) the waste disposal situation during the preparation of the plan;

(b) the nature and quantity of waste produced per year together with foreseeable future developments;

(c) the quantity of household refuse which must be disposed of, as well as the quantity of other waste which must be disposed of, either because it will be disposed of at the same time as the household refuse, or because it cannot be effectively disposed of by third parties because of the nature or quantity of the material involved;

(d) measures to be taken and financial steps required to put the plan into action.

The waste company prepares a draft of the overall plan which is then deposited with the communes, at the waste company itself and with regional development companies where it can be consulted, for a period of two months.

During this two-month period, anyone can send in written objections to the draft plan to the regional development companies or the colleges of burgomasters and aldermen involved.

Communes, authorities, establishments and businesses send their opinions, giving reasons, concerning the plan to the relevant regional development company within thirty days from the expiry of the two-month period.

The regional development companies send the objections and opinions they receive, together with their own comments, to the waste company.

The Flemish Executive sets out the plan by decree which contains details of what has been taken into consideration concerning the objections and opinions received. In the event of the Flemish Executive acting on the opinion of a regional development company, this must be done by means of a decree which must also state the reasons for such an action.

The plan is published in an extract of *Moniteur belge* and it becomes effective ten days after its publication.

The plan carries the force of a regulation.

8.6.2 The disposal of waste

8.6.2.1 THE PROHIBITION PRINCIPLE

It is prohibited to dump waste, to dispose of waste or to entrust it to third parties except in accordance with the provisions of the decree.

8.6.2.2 THE DECLARATION PROCEDURE

Producers of waste, with the exception of household refuse, are required to declare their waste to the waste company.

8.6.2.3 THE AUTHORISATION PROCEDURE

It is prohibited to create, organise, exploit, transfer, alter or convert an establishment in which waste is treated, or to change the methods used in such an establishment, without authorisation.

Six decrees issued by the Flemish Executive on 21 April 1982 control the general conditions applicable to installations intended for the disposal of waste.[33] In addition, the Flemish Executive can fix special conditions.

The permanent Deputation of the Provincial Council grants, refuses or withdraws authorisations after hearing the opinion of the college of burgomaster and aldermen of the commune where the establishment is located, in accordance with the procedure fixed by the Decree of the Flemish Executive of 21 April 1982.[34] Another decree of the same date[35] fixes the list of institutions required to provide an opinion on all authorisation applications received by the Permanent Deputation. Such an opinion must be given within thirty days of receipt of the application.

If an opinion is not given within the time stipulated, it is considered to be favourable. The Permanent Deputation of the Provincial Council can consult interested parties before proceeding further.

The Permanent Deputation of the Provincial Council decides within sixty days of receipt of the application and communicates its decision, together with the opinions received, to the waste company within ten days of taking the decision.

The waste company, the governor of the province, the college of burgomaster and aldermen, the applicant and any other interested parties can appeal to the Flemish Executive against the decision taken by the Permanent Deputation.

The appeal must be sent by registered mail within thirty days of notification from the Permanent Deputation.

The Executive makes its decision during the sixty days of the introduction of the appeal. The applicant, the waste company and the commune are notified within ten days of the decision.

The Flemish Executive can hear the views of the applicant or interested parties and can gather any opinions which may be helpful for an assessment of the appeal.

8.6.3 Prevention of the production of waste

The Flemish Executive can establish rules relating to products which:

(a) after normal use cannot be disposed of in a way which is harmless to the health of man or the environment;
(b) generate waste during their manufacture which cannot be disposed of in a way which is harmless to the health of man or the environment.

These rules can involve:

(a) the prohibition to manufacture, transport, hold, sell, offer for sale, use, acquire or offer as free or conditional gift such products;
(b) the prohibition to manufacture, transport, hold, sell, offer for sale, use, acquire or distribute such products if certain specific conditions are not complied with, including conditions relating to the product life and repairability;
(c) an obligation on the part of producers, importers or distributors to

provide such products or their packaging with a legible label or indelible inscription which is recognisable by the public.

The Flemish Executive can also establish rules relating to packaging methods and materials and to disposable products.

8.6.4 Special controls

8.6.4.1 HOUSEHOLD REFUSE

The disposal of household refuse is controlled by communal regulations; the Flemish Executive can, however, require certain specified categories of household waste to be collected separately.

8.6.4.2 INDUSTRIAL WASTE

Businesses which generate waste are responsible for the disposal of such waste, at their own expense.

Industrial waste must be dealt with by:

(a) disposal within the business which generated the waste, in accordance with the authorisation granted under the provisions of the decree; or

(b) disposal by a waste disposal organisation which must be qualified by authorisation granted in accordance with the provisions of the decree; or

(c) disposal by a person established in another Region or country who can prove that the waste will be disposed of in that Region or country.

Businesses must declare to the waste company the nature, composition and quantity of the waste, as well as the way in which it will be disposed of. The Flemish Executive can fix precise rules concerning the period in which the declaration must be made, as well as other conditions with which the declaration must comply.

The transport of industrial waste is also subject to authorisation.

8.6.4.3 SPECIAL CATEGORIES OF WASTE

Scrap cars, waste oil and dangerous waste are subject to controls similar to those relating to industrial waste.

The Flemish Executive can fix special provisions regarding the disposal of certain categories of agricultural waste.

The Executive will also issue a regulation concerning waste arising from hospitals.

8.6.5 Rates

Under article 47 of the decree, rates can be levied on the establishments which treat waste, and manufacturers, traders and users of products specified by the Flemish Executive as generating waste during their manufacture which must be disposed of, or products which must be disposed of after use.

The level of the rates to be charged is fixed by the Flemish Council. A decree of this kind has not yet been adopted.

8.6.6 Monitoring

The Flemish Executive appoints officials to monitor the implementation of the decree and its implementing provisions.

The governor of the province, the burgomaster and the officials can, in carrying out their duties:

(1) have free access at all hours of the day or night, without prior notice, to all establishments, parts of establishments, premises or workshops where waste is stored or disposed of; they may, however, only enter inhabited premises between the hours of 5.00 am. and 9.00 pm., with authorisation from the police tribunal judge;

(2) carry out any examinations, checks and enquiries and collect information they consider to be necessary to ensure that the provisions of the decree and regulations are effectively observed, and in particular they can:

    (a) interrogate individuals regarding facts to be established in the carrying out of their monitoring duties;

    (b) demand the production of books and documents required by the decree and its implementing provisions, take copies or extracts, and even remove them, against receipt;

    (c) inspect any books or documents which they consider to be necessary in carrying out their duties;

    (d) take samples, free of charge, to determine the composition of waste, and if necessary, require suitable packaging to be produced for the transport and storage of the samples;

(3) request the assistance of the communal police or gendarmerie in carrying out their duties.

In the event of an infringement, the officials have the right to prepare a written report which will be taken as fact in the absence of proof to the contrary. The person who committed the infringement must be notified of the written report within fourteen days of the infringement being ascertained. Failure to do so will invalidate the report.

8.6.7 Sanctions

Anyone who infringes or hinders the monitoring of the provisions of the Decree of 2 July 1981, or the requirements of the authorisation, is punishable by imprisonment of from eight days to one year and/or by a fine of from 100 to 100,000 francs.[36]

In the event of a repeated offence within three years following a conviction, the punishment can be increased to double the maximum.

The waste, packaging, tools and means of transport used to commit an infringement can be seized, even if they do not belong to the person who committed the infringement.

Anyone who dumps waste in contravention of the provisions of the decree is ordered by the tribunal to dispose of such waste within a specified period. The individual who committed the infringement can be required to pay the disposal costs of the commune or the waste company.

Employers are responsible under civil law for the payment of fines incurred by employees or representatives, as well as the payment of their legal costs.

Establishments subject to authorisation are required to specify the individual who will be responsible for implementing the measures required under the provisions of the authorisation.

Notes

1. *J.O.,* L 194, 25 July 1975.
2. Jean-Pierre Hannequart, Etat actuel de la réglementation des dêchets, en Belgique, *Rev. jur, ec. urban. environn,* Seres, December 1981, pp. 11–17.
3. *Moniteur belge,* 1 March 1975 and 14 February 1976.
4. *J.O.,* L 84, 31 March 1978.

5. The Commission of Approval is composed as follows:

(1) the director general of the Safety at Work Administration or his delegate, who assumes the presidency;

(2) the director general of the Hygiene and Medicine at Work Administration or his delegate;

(3) the director general of the Mines Administration or his delegate;

(4) the director general of the Pyblic Hygiene Administration or his delegate;

(5) the head of the establishment of the State Scientific Establishment, 'Institute of Hygiene and Epidemiology', or his delegate;

(6) the director general of the Agronomic Research Administration or his delegate;

(7) the official in charge of the Guarantee Fund for the Destruction of Toxic Wastes or his delegate, as soon as the fund has been constituted;

(8) three persons chosen because of their special scientitic ability;

(9) the local officials concerned belonging to the Ministries of Employment and Labour, Public Health, and Economic Affairs, competent in matters regarding establishments classed as dangerous, dirty or noxious;

(10) an official of the industrial toxicology laboratory of the Hygiene and Medicine at Work Administration;

(11) a secretary and a deputy secretary;

(12) a representative of the Regional Development Society/ies concerned.

6. Chamber of Representatives, Session 1973–1974, 14 November 1973, Doc. 684.

7. Articles 2–15, 17, 18–20 and 23.

8. He remains entirely responsible for the costs of destruction, neutralisation or disposal as well as any eventual damage caused by the toxic wastes.

9. At the time of presentation of an approval application for a centre for the destruction, neutralisation or disposal of toxic wastes, a State contribution of 10,000 francs is levied by the State to cover administration costs.

10. To this end, article 1 *bis* has been inserted in the Law of 24 February 1921, concerning the sale of poisonous, soporific, narcotic, disinfectant or antiseptic substances.

11. The nomination of a responsible person corresponds with the necessity for a person who will supervise the observation of all the legal obligations when a civil enterprise or organisation is involved.

12. It is the same protection system as that envisaged for the delegates of the Committees of Safety, Hygiene and Improvement of Work Places.

13. Article 28 of the Royal Decree of 9 February 1976.

14. Article 34 of the Royal Decree of 9 February 1976.

15. Council of State, 30 March 1971, no. 14.367, Van Limberghen.

16. Cass. 20 December 1951, *Pas.*, 1952, I, 204.

17. The Law of 26 July 1971 has transferred the attributed powers of the communal authorities with regard to the removal and treatment of wastes to the agglomerates and federations of communes. In reality, there is only one such agglomerate: the Brussels agglomerate, composed of 19 communes.

18. B. Derouaux, Propreté et embellissement du cadre de vie, in *Convention Nationale sur la qualite de la vie,* Liège, 1979, pp. 79–84.

19. *Moniteur belge*, 2–3 January 1961.

20. Royal Decree of 24 January 1969 containing measures for health policing relating to sites for the depositing of sludge and waste and the use of organic and kitchen wastes for the feeding of domestic animals, *Moniteur belge*, 12 March 1969.

21. *Moniteur belge*, 23 April 1969.

22. Royal Decree amending Chapter II, B of Heading I of the General Regulation for Protection at Work, *Moniteur belge*, 4 December 1970.

23. See also the Royal Decree of 18 July 1973, amending Heading I, Chapter II, A, of the General Regulation on Protection at Work.

152

24. See 3.2.2.
25. Article 44(1) of the law of 29 March 1962 organising land management and town planning.
26. Ministerial Circular no. 12–5 of 14 September 1964, *Moniteur belge*, 24 September 1964.
27. *Moniteur belge*, 28 August 1973.
28. See E. Bernaerts, Juridisch gedeelte. Huishoudelijke Afval, in J. C. Delacroix, M. Geeraets-Bracops, E. Bernaerts, Leefmilieu Gemeentekrediet van Belgie, Informati-dossier no. 2.
29. This applies not only to the public highway but also to streets, alleys, passages and cul-de-sacs crossing private property and joining the public highway: Cass. 29 January 1906, *Pas.*, 1906, I, 112.
30. See the final scientific report 'De Belgische Afvalbeurs—La Bourse Belge des Déchets', Programmation de la Politique Scientifique, 1980.
31. *Moniteur belge*, 25 July 1981.
32. See in particular the following decrees: Royal Decree of 30 July 1981 (establishment of statutes), *Moniteur belge*, 20 August 1981; Royal Decree of 30 July 1981 (organic framework), *Moniteur belge*, 20 August 1981; Royal Decree of 30 July 1981 (administrative statute of personnel), *Moniteur belge*, 20 August 1981; Royal Decree of 30 July 1981 (hierarchical classification of grades), *Moniteur belge*, 20 August 1981; Royal Decree of 30 July 1981 (pecuniary statute of personnel), *Moniteur belge*, 20 August 1981; Royal Decree of 14 September 1981, *Moniteur belge*, 26 September 1981.
33. *Moniteur belge*, 10 June 1982.
34. *Moniteur belge*, 15 May 1982.
35. *Moniteur belge*, 25 May 1982.
36. To be multiplied by sixty.

9
Ionising Radiation

9.1 THE OUTLINE LAW OF 19 MARCH 1958[1]

9.1.1 Introduction

The basic law is the 'Law Relating to the Protection of the Population against Dangers Arising from Ionising Radiation' of 19 March 1958.

In the application of this law:

'ionising radiation' is taken to mean radiation producing direct or indirect ionisation of matter in its path;[2]

'radioactive substances' are substances made up of elements emitting ionising radiation or containing such elements, i.e. any substance presenting the phenomenon of radioactivity.

9.1.2 Regulatory powers

The Law of 19 March 1958 gives the King extensive regulatory powers:

(a) The King is authorised to make the importation, production, manufacture, storage, transit, transport, sale, free or conditional gift, distribution or use for commercial, industrial, scientific, medical or other purposes of equipment or substances capable of emitting ionising radiation subject to conditions for the purpose of protecting the health of the population. Similarly, he can control the disposal and discharge of radioactive substances.

(b) He can establish rates to be paid to the State or approved controlling bodies to cover all or part of the administrative costs of controlling and monitoring the application of the implementing regulation of the law;[3]

(c) He is authorised to take all necessary measures with regard to producers, manufacturers, holders, transporters or users of equipment or substances capable of emitting ionising radiation in order to safeguard the population in the event of any unexpected incident endangering the health of the population. He is also authorised, in similar circumstances, to prescribe all necessary measures to prevent any danger which could result from the accidental contamination of areas, substances or products by radioactive substances.

All Royal Decrees issued in implementation of this law must be discussed by the Council of Ministers.

9.1.3 Monitoring the application of the law

The King shall appoint specially skilled agents to monitor the application of the law and its implementing decrees. National defence is, however, an important exception and the Minister of National Defence appoints agents for this task:

(1) on military property;
(2) on all other property as determined by him where equipment or substances capable of emitting ionising radiation are produced, manufactured, held or used for the requirements of the armed forces;
(3) for transport operations ordered by him of equipment or substances as mentioned above.

Without prejudice to the powers delegated to the agents of the judicial police, the appointed agents verify any infringements committed. The reports they produce are taken as fact in the absence of proof to the contrary.

They have free access at all times to factories, depots, hospitals and all establishments where equipment or substances capable of producing ionising radiation are produced, manufactured, held or used.

They can seize equipment or substances which are being produced, manufactured, held, transported or used under conditions which do not comply with the provisions of the law or its implementing decrees.

In these cases and independently of any eventual judicial prosecution, they can take all necessary measures to make safe sources of ionising radiation which could constitute a danger for the health of the population.

9.1.4 Sanctions

Article 7 punishes infringements of the provisions of the law and its implementing decrees with a fine of from 1,000 to 10,000 francs[4] and imprisonment of from three months to one year, or one of these punishments only.

9.2 THE GENERAL REGULATION FOR THE PROTECTION OF THE POPULATION AND WORKERS AGAINST THE DANGER OF IONISING RADIATION

9.2.1 Introduction

On 28 February 1963 the royal decree acting as a General Regulation for the protection of the population and workers against the danger of ionising radiation was issued as a result of the outline Law of 19 March 1958 and in implementation of the directives of the Council of the European Atomic Energy Community of 2 February 1959 fixing the basic provisions on the health protection of the population and workers against dangers resulting from ionising radiation.

9.2.2 Application

This regulation applies to:

(1) the importation, production, manufacture, storage, transport and use for commercial, industrial, scientific, medical or other purposes of equipment, installations or substances capable of emitting ionising radiation;

(2) the offer for sale, sale, free or conditional gift of substances capable of emitting ionising radiation, or equipment or installations containing such substances;

(3) the treatment, handling, storage, disposal and discharge of radioactive substances and waste.

The regulation does not apply to:

(a) television sets;

156

(b) equipment or installations which can only emit ultraviolet radiation;
(c) military equipment or installations;
(d) the transport of equipment or substances capable of emitting ionising radiation ordered by the Minister of National Defence.

It does not concern natural radiation and the various forms of ionising radiation arising from natural terrestrial or cosmic sources.

9.2.3 The control of classified establishments[5]

Establishments which use or hold radioactive substances are classified into four groups.[6]

Establishments in Classes I, II and III are subject to prior authorisation. Class IV establishments are not subject to any authorisation or declaration formality; they are, of course, required to observe other provisions of the General Regulation.

9.2.3.1 CLASS I ESTABLISHMENTS

Authorisation for Class I establishments is granted by the King.

The application is sent to the governor of the province in which the establishment is located. This application must include information and documents as precisely listed in article 6(2). The application must also be accompanied by a note indicating the proposed measures for the organisation, purification and disposal of any liquid, solid or gaseous radioactive waste.

The governor sends a copy of the application to the burgomaster of the commune, who will post up a notice at the site where the establishment is located and at the town hall; this notice must state the purpose of the application and the possibility of making objections or comments during the fifteen days following the first day that the notice was posted up at the town hall.[7]

The application, together with any objections and comments received, are submitted first to the college of the burgomaster and aldermen and then to the deputed states of the province. Finally, the complete file is passed to the 'Special Commission' which plays a key role in the authorisation procedure.[8]

This Special Commission can collect additional information on the security and salubrity of the establishment. In instances covered by article 37 of the Euratom Treaty, the Special Commission can request the opinion of the Commission of the European Communities. It can also consult the Commission on general and specific aspects of the safety and salubrity of the establishment. Equally, it can summon the applicant for further discussion.

Subsequently the Special Commission issues a provisional opinion which is sent to the applicant by registered post. The communication of the provisional opinion is an important formality and if not carried out the opinion is invalidated.[9] The applicant has thirty days in which to make any comments.

Finally, the Special Commission reconsiders the case and issues a final reasoned opinion.

In the event of a favourable opinion, special conditions can be imposed which the Commission considers necessary to ensure the safety and salubrity of the establishment, that are not covered by the General Regulation.

If the Commission issues an unfavourable opinion, the authorisation is refused.

When the Commission's opinion is favourable, any subsequent decree refusing authorisation must be justified.

Authorisation is granted by royal decree countersigned by the Minister of Employment and Labour and the Minister of Public Health.

There is no administrative appeal but a judicial appeal can be made to the Council of State.

9.2.3.2 CLASS II ESTABLISHMENTS

These are subject to prior authorisation granted by the deputed states of the provincial council of the province in which the establishment is located.

The authorisation application is sent to the governor of the province. The information and documents which must accompany the application are broadly the same as those required for Class I applications.

The governor sends a copy of the application to the burgomaster of the commune involved who submits it to the opinion of the College of Burgomaster and Aldermen; within thirty days the application and opinion must be returned. There is no public inquiry or publication of the authorisation application for Class II establishments.

The file is then passed to the Provincial Consultative Committee.[10]

This Committee proceeds in the same way as the Special Commission involved with Class I authorisation applications. Finally, the governor submits the application and opinions received to the deputed states of the provincial council, who must issue a decision by decree within one month.

The decree must contain the conditions proposed by the Provincial Consultative Committee.

When the opinion of the Provincial Consultative Committee is favourable, any subsequent decree of refusal must be justified.

The decision is published by posting up a notice at the town hall and at the site where the establishment is located. This notice must mention that interested parties can see the decision at the town hall and that a means of appeal is open to them.

Appeals must be sent to the King within fifteen days from the posting up of the notice at the site where the establishment is located. The appeal is sent to the Special Commission which acts in the same way as for applications for Class I establishments.

The royal decree on the outcome of the appeal is countersigned by the Minister of Employment and Labour and the Minister of Public Health.

In the event of an authorisation being refused despite a favourable opinion, the decree must give reasons for refusal.

9.2.3.3 CLASS III ESTABLISHMENTS

These establishments must make prior declaration to the governor of the province with a view to obtaining authorisation. The governor submits the declaration to the deputed states of the provincial council which grant or refuse authorisation within fifteen days.

The deputed states must judge whether or not the applicant is sufficiently organised and equipped to protect the workers and population effectively against the dangers involved in the nuclear activities of his business.[11]

The Provincial Consultative Committee is responsible for distributing the lists containing the conditions of use which apply to various types of Class III establishment. The deputed states are required to impose conditions relating to the type of establishment in question in the authorisation decree.

An administrative appeal is open to the applicant through the burgomaster of the commune in which the establishment is located, and various officials.

The appeal is sent to the Special Commission which issues its opinion on the decision of the deputed states in the same way as for applications for Class I establishments.

Contrary to the procedure for Class II establishments, the King, who takes the final decision on the appeal, can grant an authorisation in spite of an unfavourable opinion from the Special Commission or, alternatively, can refuse authorisation despite a favourable opinion, without giving his reasons for refusal.

9.2.3.4 MIXED ESTABLISHMENTS

Authorisation applications relating to installations including establishments belonging to several classes are treated in accordance with the provisions relating to the highest class involved.

Establishments classed as dangerous, dirty or noxious, in the meaning of heading I of the General Regulation for the protection of labour, and steam equipment subject to the provisions of heading IV of this regulation, which are attached to a nuclear establishment and form an essential part of its function or exploitation, can only be authorised by the authority responsible for nuclear establishments. The information and documents to be supplied are those defined by the General Regulation for protection of labour. The inquiry formalities are those defined for the nuclear business under consideration.

9.2.3.5 ESTABLISHMENTS OPERATED BY THE STATE

The authorisation procedure also applies to establishments operated by the State or by a public corporation.

However, with regard to Class II establishments, it is the 'Special Commission' which must give an opinion, instead of the Provincial Consultative Committee and the authorisation is granted by royal decree; with regard to Class III establishments the authorisation is granted by the Minister of Public Health.

9.2.3.6 PROVISIONS COMMON TO CLASS I, II AND III ESTABLISHMENTS

The General Regulation states that managers or directors are required to respect the conditions in authorisation decrees. The authority responsible must be notified without delay by registered post of any change arising in the appointment of the director or manager responsible for the establishment.

The authority responsible can exempt the authorisation applicant from supplying certain information or documents. The authorisation decree can fix further means of communicating such information and documents.

Authorisations can be granted for unlimited duration or for a specified period. They cannot be granted for a trial period.

Authorisations can be transferred from one operator to another on condition that the authority which granted the authorisation is notified of the assignment without delay by registered post.

9.2.3.7 EXEMPTION FROM AUTHORISATION

Establishments where occasional trials and tests are carried out on materials or processes involving the use of ionising radiation are not considered as classified establishments, and are therefore exempt from prior authorisation provided that:

(a) the operations are carried out exclusively by personnel from a third establishment duly jointly authorised to undertake such operations by the Minister of Employment and Labour and the Minister of Public Health;
(b) the approved organisation responsible for controlling the establishment or the approved expert directing the physical control service approves the operation.

9.2.3.8 APPROVAL OF THE INSTALLATION OF CLASS I AND II ESTABLISHMENTS

The authorisation granted to Class I and II establishments enables the applicant to undertake construction work on his own responsibility and to proceed with installations in accordance with the terms of the authorisation granted.

He does, however, also require an operating authorisation. The safety system of the installations must be passed by an organisation approved by the Minister of Employment and Labour and by the Minister of Public Health.[12] Approval depends on compliance with the general and special conditions of the authorisation. Installations can only be started up if the report of the organisation involved is entirely favourable and formally authorises the starting up operation.

The director or manager responsible for the establishment is required to inform the officials responsible for monitoring, of the anticipated starting date, at least fifteen days in advance by registered post.

9.2.3.9 REQUIREMENT OF 'TOWN PLANNING' PERMIT

In its judgement no. 18.852 of 19 March 1978, the Council of State declares that article 15 of the Royal Decree of 28 February 1963—which provides for prior authorisation to enable the applicant to undertake construction work on his own responsibility and to proceed with installations in accordance with the authorisation granted—does not make the implementation of town planning law subject to the granting of the authorisation required by the implementing decree of the law on the protection of the population.[13] In other words, the builder of a nuclear establishment must obtain not only a 'nuclear' permit, but also a 'town planning' permit.

9.2.3.10 FURTHER EVENTS

The General Regulation contains a series of provisions concerning events which may occur after the authorised establishment has been brought into service.

(a) All plans for major alterations or extensions to the establishment are subject to new authorisation and an application must be submitted to the responsible auhtority.

If the alteration or extension leads to a change in classification from a lower class to a higher one, the authorisation procedure which applies is the procedure relating to the higher class involved.

(b) The responsible authority can make additions or alterations to the authorisation decree on the suggestion of:

(1) the Special Commission with regard to establishments authorised by royal or ministerial decree;

(2) the Provincial Consultative Committee with regard to establishments authorised in the last instance by the deputed states.

The Special Commission and the Provincial Consultative Committees can act on the initiative or suggestion of one of the officials responsible for monitoring activities.

An appeal is open to interested parties for establishments authorised by the deputed states. This appeal suspends the decision under attack.

(c) When an establishment does not comply with the provisions of the General Regulation or the conditions of the authorisation decree, the authority responsible can suspend or withdraw the authorisation after obtaining the opinion of the Special Commission and the Provincial Consultative Committee.

An appeal to the King is open with regard to establishments authorised by the deputed states. This appeal suspends a decree suspending authorisation but not a decree withdrawing authorisation.

(d) In the event of operations ceasing, for whatever cause, in a Class I or II establishment which holds radioactive substances or equipment, the enterprise or, where applicable, the individual legally appointed to liquidate the business, is required to advise the Minister of Public Health immediately. They must also advise the authorities designated in article 78 of the General Regulation.[14]

Such substances or equipment must be sent to a destination which guarantees their disposal or re-use under satisfactory conditions.

9.2.4 Organisation of general protection

9.2.4.1

Chapter III of the General Regulation consists of a collection of provisions for the protection not only of workers but also of third parties.

9.2.4.2

Section I contains basic requirements concerning maximum permitted dose level, exposure and contamination. It should be noted that no worker under eighteen years of age can carry out an activity during which he will be occupationally exposed to the risk of ionising radiation. This also applies to pregnant and nursing mothers.

9.2.4.3

Section II deals with physical and medical checks. The head of the business is required to organise a service to undertake *physical checks* for health protection which is in charge of organising supervision of measures necessary to ensure compliance with the provisions of the regulation concerning safety and hygiene at work, and the safety and salubrity of the neighbourhood, with the exception of provisions relating to medical checks.

The *medical examination* of workers occupationally exposed to ionising radiation is carried out in accordance with the provisions of Heading II,

Chapter III, Section I of the General Regulation for the protection of labour, or Chapter II of the Regent's Decree of 25 September 1947 acting as a General Regulation on health and hygiene measures for underground quarry and mine workers, depending on the case under consideration.

9.2.4.4

Section III sets out in detail the necessary general protective devices and processes.

The basic principle is unequivocal: businesses are required to make use of all efficient protective devices necessary to comply with the maximum permitted doses. In principle these devices are based on a weekly dose of 0.1 rem.

It is forbidden to enter or remain in a controlled zone without authorisation from the head of the business or his deputy. This authorisation can only be granted for professional or operational reasons. Persons permitted to enter these zones are listed in an 'ad hoc' register which includes their identity and, where appropriate, the purpose of their visit.

The radioactivity symbol must be displayed:

(a) at each entrance to the controlled zone;
(b) on the external doors and walls of rooms where one or more sources of ionising radiation are in operation, held or stored;
(c) on vehicles, casks and receptacles in which such sources are transported or contained;
(d) on all equipment emitting ionising radiation with the exception of equipment liable to be used in Class IV establishments.

Any additional information warning persons at risk of the dangers they could be running must be displayed visibly and legibly underneath the radioactivity symbol.

9.2.4.5

Section IV controls the collection, treatment and disposal of *radioactive wastes* with a radiation level exceeding the natural level of radiation.[16]

(a) The discharge of *liquid radioactive* waste is forbidden:

into surface waters, including open air drainage canals, lakes, ponds, river basins and stretches of water when the radioactive nuclide content of the waste exceeds one tenth of the maximum permitted content for drinking water contained in the annex of the General Regulation;

into underground drains and sewers when the radioactive nuclide content exceeds the maximum permitted concentration;

into the ground when the radioactive nuclide content exceeds one-hundredth of the maximum permitted concentration.

Liquid wastes with a radioactive nuclide content exceeding the specified limits or conditions fixed in the authorisation and which can therefore not be discharged must be stored in water-tight receptacles ensuring sufficient protection as a result of their treatment or eventual discharge after a reduction in radioactivity, or after dilution, to meet the fixed limits; this last process can only be used if it is explicitly permitted in the authorisation decree.

(b) The discharge of *gaseous radioactive* waste into the atmosphere in the form of gas, dust, smoke or vapour is forbidden when the concentration of radioactive substances at the point of emission into the atmosphere is higher than one-tenth of the permitted maximum concentration for inhaled air in the annex to the General Regulation.

The authorisation decrees granted to Class I and II establishments can exempt them from this provision; in that case, the instantaneous and average maximum permitted concentrations of radioactive nuclide liable to be contained in the gaseous effluents are fixed by the authorisation decree.

Gaseous effluents with a radioactive nuclide content exceeding the limits or conditions fixed in the authorisation must be filtered, stored or treated so that the radioactive nuclide concentration of the air emitted into the atmosphere is less than the permitted limit.

(c) *Solid radioactive* wastes are 'carefully' collected in water-tight receptacles ensuring sufficient protection; they are subsequently treated and then concentrated in such a manner as to avoid any risk of dispersion of radioactivity under any form.

The discharge of solid radioactive waste into surface waters, drains and sewers is forbidden.

(d) Liquid and solid radioactive wastes which cannot be discharged are contained and stored in closed, solid, water-tight receptacles which are kept in fireproof surroundings in order to prevent any dispersion of radioactive substances.

The deposit on or into the ground of these wastes is forbidden. Authorisation decrees granted to Class I and II establishments can exempt them from this prohibition provided that the wastes are contained in water-tight receptacles which are resistant to physical and chemical agents and on condition that they are combined with inert, insoluble substances in such a manner as to make the dispersion or escape of radioactive matter impossible or reduced to such a low level that it does not present a danger.

A permanent inventory is kept, as accurately as possible, of radioactive wastes contained in storage and discharges carried out. This inventory is kept available at all times for inspection by officials in charge of monitoring.

In fact, since 1967 Belgium has carried out nine discharge operations at sea; they are carried out within the context of the mechanism instituted by the decision of the OECD Council of 22 July 1977. These operations have enabled 15,164 tonnes of radioactive matter to be disposed of.

> The last dumping programme, which dates from 22 June 1979, was jointly organised by the Netherlands, Belgium and Switzerland. 872 tonnes from Belgium and 412 tonnes from Switzerland were loaded on to a Dutch ship at the port of Zeebrugge. The Belgian waste came from CEN at Mol where low-level radioactive wastes are centralised, treated, conditioned and packed. The operation was monitored by the radiation control service of CEN, a delegate from the Minister of Public Health and two OECD delegates. The dumping of the wastes was carried out in accordance with the London Convention, the IAEA requirements and the OECD recommendations. An evaluation of the discharge site was carried out in 1979 and was favourable.[17]

9.2.5 Importation, transit and distribution of radioactive substances

9.2.5.1

The importation and transit of radioactive substances or equipment containing radioactive substances excluding those in Class IV can only be carried out by persons and businesses which have obtained authorisation to do so from the Minister of Public Health.

The authorisation is granted for a limited period; it can be general or specific.

The authorisation act can impose special conditions which are not contained in the general regulation.

The authorisation can be withdrawn at any time by justified decision from the Minister of Public Health.

The transit authorisation is subject to an undertaking by a person or organisation established in Belgium and approved for the purpose by the authorisation act, to directly and wholly indemnify the victims of any accident. This undertaking is not required if the transport comes under the Convention of Paris on civil liability in the field of nuclear energy, its supplementary convention and additional protocols approved by the Law of 1 August 1966.

In waiving these provisions, the Royal Decree of 27 July 1966 recognises the validity on Belgian territory of authorisations concerning Belgian territory for the importation and transit of radioactive substances or equipment containing radioactive substances granted by the administration responsible in the Netherlands or the Grand Duchy of Luxembourg.[18]

A Ministerial Decree of 9 May 1977 appoints the customs offices through which the importation of radioactive substances or equipment containing radioactive substances subject to authorisation can be carried out.

9.2.5.2

The *distribution* of radioactive substances or equipment containing radioactive substances can only be made to authorised establishments.[20]

9.2.6 Transport of radioactive substances

The transport of radioactive substances, whatever the method of transport used, including private vehicles, must be carried out in accordance with the provisions of the international conventions in force; in addition, prior authorisation is required from the Minister of Public Health, except for radioactive substances which do not exceed a level specified in the General Regulation.

This authorisation can be general, specific or special.

A general authorisation can be granted to the transporter who wishes to transport radioactive substances regularly. It can be granted for a period of not more than five years. It is renewable on request by the transporter.

A specific authorisation can be granted to the transporter who wishes to transport such substances occasionally.

A special authorisation is required for types of transport defined in article 57 of the General Regulation.

The authorisation can fix specific conditions relating to the transport, and can even require the presence of an escort or convoy, particularly for transport subject to special authorisation.

The general, specific or special authorisation can be refused or withdrawn at any moment by justified decision from the Minister of Public Health. A new application can be presented if the circumstances which led to the refusal or withdrawal of the authorisation have disappeared or altered.

The Royal Decree of 27 July 1966 recognises the validity of authorisations concerning Belgian territory for the transport of radioactive substances or equipment containing radioactive substances granted by the responsible administration in the Netherlands or the Grand Duchy of Luxembourg.

If, during the transport of radioactive matter, it appears that a danger threatens the safety of the population, the person in charge of the transport is required immediately to inform the first aid centre or, where this is not possible, the nearest police station, police commissioner or fire service.

The local health inspector and the Institute of Hygiene and Epidemiology are informed as quickly as possible either by the person in charge of the transport or by the services mentioned above.

Notification of an accident does not exempt the transporter from taking all necessary on-the-spot protection measures required under the circumstances.

9.2.7 Nuclear propulsion

9.2.7.1

The construction of a ship or vehicle propelled by nuclear energy is subject to prior authorisation granted by royal decree.

The information and documents which must accompany the authorisation application are the same as those required for Class I establishment authorisation applications.

It is necessary to obtain the opinion of the College of Burgomaster and Aldermen, the deputed states and the Special Commission. In the event of an unfavourable opinion from the Special Commission, the authorisation is refused.

Ships or vehicles propelled by nuclear energy are also subject to obligatory approval.

9.2.7.2

The movement and parking of vehicles and ships propelled by nuclear energy in territorial waters, national airspace or on Belgian soil are subject to prior authorisation from the Minister of Public Health, who obtains the opinion of the Minister of Communications. This authorisation can involve special conditions with regard to the convoy, route, docking, landing, parking, duration of parking and security.

9.2.8 Ionising radiation in human or veterinary medicine

9.2.8.1

In order to ensure the safety of products used in human or veterinary medicine, no one can import, manufacture, prepare, offer for sale or sell radioisotopes used in unsealed form in human or veterinary medicine without first obtaining authorisation from the Minister of Public Health.

This authorisation specifies the substances and preparations for which it is valid, as well as the location where the operations can be carried out.

9.2.8.2

Equipment used for diagnostic purposes (radioscopy, radiography, dental radiography) and equipment used for therapeutic purposes (radiotherapy, curietherapy, gammatherapy, therapy by particle accelerator) must be of a type approved by the Minister of Public Health or an official of the Public Hygiene Administration appointed by him.

The equipment for diagnostic or therapeutic purposes mentioned above, in addition to radioisotopes used as such in unsealed form for diagnostic or therapeutic purposes, or in sealed form for therapeutic purposes, can only be used by qualified personnel defined in the General Regulation who must be authorised to this effect by the Minister of Public Health.

9.2.9 Prohibitions and authorisations

It is forbidden:

(a) to use equipment emitting ionising radiation in the footwear trade; the importation, storage and transport of such products and equipment is also forbidden;
(b) to add radioactive substances to foods, beauty products, cosmetics, toys or products and objects for domestic use;
(c) to treat foods or medicines with ionising radiation.[22]

It is forbidden to use radioactive substances and equipment or installations capable of emitting ionising radiation for agricultural, livestock or entomological research except in specially licensed premises, in order to prevent danger to man and animals.

The Minister of Public Health can, if necessary, prohibit the distribution of certain radioactive substances.

On the other hand, however, if the Superior Council of Public Hygiene is in favour, and providing the conditions it imposes are observed, he can authorise certain prohibited activities for research purposes, for the sterilisation of materials used for medical or surgical purposes, etc.

9.2.10 Exceptional measures

Chapter X of the General Regulation (articles 66–69) sets out measures to prevent the theft or loss of radioactive substances and measures concerning accidents, accidental irradiation and decontamination.

9.2.11 Monitoring of the national territory and of the population as a whole

The Minister of Public Health is responsible for monitoring radioactivity throughout the territory.

In normal conditions this involves:

(1) the regular measurement of radioactivity of air, water, soil and the food chain;
(2) the evaluation and monitoring of radiation doses received by the population.

The Institute of Hygiene and Epidemiology monitors the population as a whole on the basis of the measurements carried out and the information obtained relating to the doses received by individuals who are occupationally exposed to ionising radiation.

The Minister of the Interior, in his civil defence capacity and with the collaboration of the Minister of Public Health, takes care of the preparation and implementation of measures to be taken in the event of accident or abnormal increase in the ambient level of radioactivity.

9.2.12 The approval of experts, organisations and doctors

The General Regulation provides for the approval of experts, organisations and doctors.

There are three classes of *expert*:

Class I experts are those who can direct the physical examination service in Class I establishments or nuclear propelled vehicles and those who can carry out inspection visits in such establishments and vehicles.

Class II experts are those who can direct the physical examination service in Class II establishments and those who can carry out inspection visits in such establishments.

Class III experts are those who can direct the physical examination service in Class III establishments and those who can carry out inspection visits in such establishments.

There are two classes of *organisation:*

Class I organisations are those in charge of duties regarding Class I, II and III establishments and nuclear propelled vehicles.

Class II organisations are those in charge of duties inside Class II and III establishments.

These organisations must be constituted in the form of non-profit-making associations.

The organisation, its director and experts cannot be the builder, supplier or installer of the equipment or installations under their control, nor their agent.

They cannot be involved in trading, representing or monitoring the equipment or installations, nor can they be the agent of persons involved in such trade, representation or maintenance.

They cannot trade in substances capable of emitting ionising radiation, nor dispose of waste.

Approved organisations must set up an internal monitoring commission which must be independent from the statutory bodies.

Each monitoring commission is made up of:

one president;

the director of the organisation, or when he is prevented from attending, his deputy;

three effective members and three substitute members representing employers;

three effective and three substitute members representing workers.

The ministers responsible appoint the president from the officials of the Safety at Work Administration.

The approval of experts and organisations can be restricted:

(a)   territorially;
(b)   in time;
(c)   with regard to the nature of the equipment and installations under their control.

The medical inspections provided for in the regulation are carried out by medical, surgical or obstetrics doctors previously approved by the Minister of Public Health.

The list of approved doctors is published in *Moniteur belge*.

9.2.13 Monitoring[23]

9.2.13.1

Monitoring the application of the provisions of the General Regulation is ensured by the burgomasters who are responsible for making sure that Class I, II and III establishments as well as construction yards for nuclear driven ships and vehicles are properly authorised, by the officials designated in article 78 of the General Regulation,[24] and by the officials of the Protection against Ionising Radiation Service created in the Ministry of Public Health by the Royal Decree of 14 August 1981.[25]

9.2.13.2

The burgomasters proceed with the closure of unauthorised establishments. If they discover a situation which endangers the safety of the workforce or the population, they take suitable measures to prevent such danger occurring.

They immediately inform the technical officials designated in the General Regulation and the Civil Defence Administration. If necessary, they order the evacuation of the workforce, the public or the neighbourhood.

They take similar action if one of the officials appointed under article 78 informs them of a situation which could endanger the health or safety of the workforce or population.

In both cases the head of the business can appeal to the King. This appeal does not suspend the intervening decision.

9.2.13.3

The Council of State has declared that the General Regulation constitutes an autonomous control which does not exclude the power of the communal council to issue regulations on the subject:

> in that protection against the danger of ionising radiation cannot be considered as a proper function under article 50 of the Decree of 14 December 1789 relating to the constitution of the municipalities, or as a subject of control under the care and authority of the municipal bodies mentioned in article 3 of the Decree of 16–24 August 1790 on the organisation of the courts; the task of a communal body which participates in the implementation of this regulation is merely a task of collaboration with a service which does not come under the responsibility of local power but comes under the authority of the government. When the King cancels an order issued in the execution of a duty for which the burgomaster is responsible, the latter cannot justifiably claim injury to the prerogatives of his function as a body with local power in order to bring the Royal Decree of cancellation before the Council of State.[25]

9.2.13.4

The burgomasters and other officials and agents responsible for monitoring can order the sealing off or locking up of equipment or substances capable of emitting ionising radiation. They can take measures to render dangerous sources harmless.

When the seizure of substances or objects capable of emitting ionising radiation is under dispute and the jurisdiction responsible is not in a position to make a decision within a period compatible with the health and safety of the population, these substances or objects, on the order of the King's Procurator and according to the instructions of the technical officials of the Public Hygiene Administration, are either placed in storage or considered to be radioactive waste and treated as such.

9.2.14 The public company for the management of radioactive waste and fissile materials

Article 179, ss.2 of the Law of 8 August 1980 concerning budgetary proposals for 1979–80 created a public company in charge of storage of treated radioactive wastes, disposal of wastes, transport of radioactive wastes and enriched fissile materials, as well as the storage of plutonium.

The basic reason for the creation of this undertaking was:

> to ensure, without pursuit of gain, the most rational, co-ordinated and uniform management possible of all operations relating to nuclear materials. These operations cover transport, storage and disposal, and eventually, treatment of these materials outside the installations of the producers and users; it is necessary that all measures are taken in order to prevent exposure—at short range as well as long range—of the population or the environment to the risks which are inherent in these materials.

> The legislature therefore wishes that management to be entrusted in one single undertaking under public control, in order to guarantee that the public interest shall prevail in all decisions to be taken in this context.[27]

The working rules for the company were established by Royal Decree of 30 March 1981.

The company has a Board of Directors of twenty members, all representing public authorities. It is placed under the administrative control of the Minister of Economic Affairs.

The company does not substitute for the authorities designated in the General Regulation for the Protection of the Population against the Dangers of Ionising Radiation, or in the law concerning State security in the field of nuclear energy.

9.3 THE LAW OF 4 AUGUST 1955 CONCERNING STATE SECURITY IN THE FIELD OF NUCLEAR ENERGY

This outline law enables the King to determine the security measures necessary in the interests of national defence and State security, to which research, nuclear materials and production methods carried out or used by institutions and establishments and actual or legal persons who have access to information, documents or matters which they have obtained either directly from the government or with its authorisation must be subject.

These security measures are aimed at the organisation, safeguarding and surveillance of all premises used for research or works, the safe-keeping of documents or matter, as well as the identification of conditions required to gain access to such premises or to carry out an activity therein.

9.4 CIVIL RESPONSIBILITY

9.4.1

A Law of 9 August 1963 defines the civil responsibility of the operator of a nuclear ship. The operator of a nuclear ship is objectively responsible for all damage proved to be the result of a nuclear accident caused by the ship (strict liability).

No person other than the operator of a nuclear ship is responsible for such damage.

The responsibility of the operator of a nuclear ship is limited to 5,000 million Belgian francs for a single nuclear accident. This sum includes legal costs.

The King can increase this limit on the debated opinion of the Council of Ministers.

Organisations which provide financial support or assistance such as medical care, in accordance with the laws and regulations, can act for the victims of nuclear accidents or their representatives in relation to the operator of a nuclear ship in order to obtain their rights, until the sums paid in financial support or sums representing unpaid costs are recovered.

Only the civil tribunal of Antwerp has competance to decide in the first resort on actions based on the provisions of this law and the international conventions on the subject.

9.4.2

Belgium has ratified and approved the Convention on Civil Liability in the Field of Nuclear Energy, signed in Paris on 29 July 1960, its supplementary Convention of 1963 and additional protocols.

The Law of 18 July 1966 established certain measures for immediate application.[28] A Royal Decree of 13 May 1980 fixes the maximum liability on the part of the operator at one billion Belgian francs for damage caused by a nuclear accident.

9.5 THE INTERMINISTERIAL COMMISSION ON NUCLEAR SAFETY AND STATE SECURITY IN THE NUCLEAR FIELD

A Royal Decree of 15 October 1979 created an 'Interministerial Commission on Nuclear Safety and State Security in the Nuclear Field', with the task of researching ways of ensuring the protection of the workforce and population against dangers which could result from activities involving the use, processing, transport or storage of radioactive substances either outside or inside installations where such activities are carried out.

The Commission establishes a co-ordination plan for the activities of all the ministerial departments concerned. It informs the Ministerial Environment Committee of the implementation of this plan. The Commission consists of:

(1) a president elected by the Minister of Public Health and the Environment;
(2) a vice-president elected by the Minister of Employment and Labour;
(3) a member elected by the Minister of Employment and Labour;
(4) a member elected by the Minister of Public Health and the Environment;
(5) a member elected by the Minister of the Justice;
(6) a member elected by the Minister of the Interior;
(7) a member elected by the Minister of National Defence;

(8) a member elected by the Minister of Economic Affairs;
(9) a member elected by the Minister of Foreign Affairs;
(10) the president of the Special Commission on Ionising Radiation.

9.6 THE NON-PROLIFERATION OF NUCLEAR ARMS

The Law of 14 March 1975 approved the Treaty on the Non-proliferation of Nuclear Arms, signed in London, Moscow and Washington on 1 July 1968 and the Agreement between the Kingdom of Belgium, the Kingdom of Denmark, the Federal Republic of Germany, Ireland, the Republic of Italy, the Grand Duchy of Luxembourg, the Kingdom of the Netherlands, the European Community of Atomic Energy and the International Atomic Energy Agency in implementation of paragraphs 1 and 4 of article III of the Treaty on the Non-proliferation of Nuclear Arms signed in Brussels on 5 April 1974.[29]

This Agreement involves a series of obligations which, for the most part, fall on the European Community as such, but which, in certain cases, fall indirectly on each Member State party to the Agreement. It is for this reason that each of these States, including Belgium, has the obligation to accept the inspections and verifications that the International Atomic Energy Agency must carry out under the terms of the above Agreement, or have such inspections and verifications accepted by the persons or businesses which, in whatever manner, produce, use or store on their territory, raw materials or special fissionable materials.

The Law of 20 July 1979 contains proper provisions to empower the International Atomic Energy Agency to carry out these inspections and verifications on Belgian soil.[30]

ANNEX I

Establishments are classified into the following classes:

Class I

(1) Nuclear reactors.
(2) Establishments where fissionable materials (except natural uranium) are used or stored in quantities greater than half their minimum critical mass.

(3) Factories for the extraction of enriched or non-enriched irradiated nuclear fuels.

Class II

(1) Establishments where quantities of fissionable substances not included in Class I (excluding natural uranium) are used or stored.
(2) Establishments where quantities of radioactive nuclides with a total activity within a range of values specified in an attached table (including natural uranium) are used or stored.
(3) Establishments engaged in the collection, treatment, processing and storage of radioactive wastes.
(4) Establishments using non-transportable equipment generating X-rays capable of functioning at a peak tension of more than 200 kV.
(5) Particle accelerators.

Class III

(1a) Establishments where quantities of radioactive nuclides with a total activity included in a range of values specified in an attached table are used or stored.
(1b) Establishments where sealed sources containing quantities of radioactive nuclides with a total activity exceeding maximum values given in the table, but not more than one hundred times these values, are used or stored. When sealed sources containing tritium or radioactive isotopes of argon, krypton or xenon are involved, the Minister of Labour and Employment and the Minister of Public Health can classify establishments where quantities of radioactive material with a total activity exceeding the values mentioned in the above paragraph are stored or used in Class III.
(2) Establishments using non-transportable equipment generating X-rays which can function at a peak tension equal to or less than 200 kV.
(3) Establishments using transportable equipment generating X-rays.

Class IV

(1) Establishments where quantities of radioactive nuclides with a total activity within a range of values given in an attached table are used or stored.
(2) Establishments holding or using equipment containing radioactive substances exceeding the values fixed in ss.1 of this class, on condition that:
    (a) these radioactive substances are effectively protected against all contact and leakage;

(b) the dose does not exceed 0.1 millirems per hour at all accessible points at a distance of 0.1 meters from the surface of the equipment;

(c) the equipment is of a type approved by the Minister of Labour and Employment and by the Minister of Public Health.[31]

(3) Establishments where radioactive substances in whatsoever quantity with a concentration of less than 0.002 microcuries per gram, and with regard to natural solid radioactive substances, of 0.01 microcuries per gram, are used or stored.

Notes

1. *Moniteur belge*, 30 April 1958.
2. That is, electromagnetic radiation (X-rays or gamma-rays) and corpuscular radiation (electrons, positrons, beta rays, protons, neutrons, alpha particles, heavy particles etc.).
3. See the Royal Decree of 22 March 1974, *Moniteur belge*.
4. To be multiplied by sixty.
5. See J. M. Didier, Procédure d'autorisation pour la construction et l'exploitation d'établissements nucléaires dans l'Etats membres de la communité Européene, *EUR* 5284, 1976, pp. 4–24; Mary Sancy, Procédure d'obtention des permis de bâtir et des permis d'exploitation des établissements nucléaires en Belgique, *Rev. jur. econ. urban. environnem.*, Seres, December 1977, no. 2, pp. 25–29.
6. See Annex I of this chapter.
7. It is unnecessary to state that the fifteen days envisaged for the public inquiry is not sufficient for effective participation on the part of the population. In addition, unlike the situation with regard to the control of classified establishments, no public session is held during which comments and observations can be made on nuclear establishments.
8. The 'Special Commission' is made up as follows:

    (a) the Director General of the Public Hygiene Administration or his deputy;
    (b) the Chief Inspector of Public Hygiene or his deputy;
    (c) the Director of the Institute of Hygiene and Epidemiology;
    (d) the Director General of the Safety at Work Administration or his deputy;
    (e) tme Chief Engineer of the Technical Inspectorate of Work, the Head of the District in question, or his deputy;
    (f) the Director General of the Administration of Hygiene and Medicine at Work or his deputy;
    (g) the Atomic Energy Commissioner or his deputy;
    (h) six persons elected on the basis of their scientific qualifications, especially in the following disciplines: nuclear physics, nuclear chemistry, radiobiology, radioprotection, safety and technology of nuclear installations, metallurgy, meteorology, geology and hydrology. These persons are jointly appointed by the Minister of Employment and Work and the Minister of Public Health;
    (i) a secretary and assistant secretary jointly appointed by the Minister of Employment and Work and the Minister of Public Health.

9. See A. Adriaens, Bescherming van de bevolking en van de werknemers tegen het gevaar van ioniserende stralingen, *Adm. Lex.*, 1965, p. 18.
10. The 'Provincial Consultative Committee' is made up as follows:

    (a) the Inspector in Chief of Public Hygiene or his deputy;
    (b) the inspector of Hygiene for the province;

     (c)   the Chief Engineer of the Technical Inspectorate of Work, chief of the district in question, or his deputy;

     (d)   the Engineer of the Technical Inspectorate of Work of the province;

     (e)   a delegate from the Mines Administration, if the establishment is monitored by that Administration;

     (f)   a secretary or assistant secretary jointly appointed by the Minister of Employment and Work and the Minister of Public Health.

11. A. Adriaens, *op. cit.*, p. 30.

12. See 9.2.12.

13. *Adm.Publ.*, June 1979, no. T3, p. 223, with the opinion and report of M. Dumont.

14. See note 24.

16. See Mary Sancy, La Belgique et les déchets radioactifs, *Rev.jur.urban.envir.*, Seres, April 1979, pp. 7–9.

17. Mary Sancy, L'immersion des déchets radioactifs, *Rev. jur, econ, urban, envir.*, Seres, June 1980, p. 19. It should be noted that Belgium has not yet ratified the London Convention of 29 December 1972 on the prevention of pollution of the sea resulting from the dumping of wastes.

18. *Moniteur belge*, 30 August 1966.

19. *Moniteur belge*, 4 August 1977.

20. Except for exemptions concerning Class IV establishments.

21. See Mary Sancy, La Belgique et les déchets radioactifs, *op. cit.*, p. 9–11.

22. See the Ministerial Decree of 10 July 1980, *Moniteur belge*, 19 August 1980.

23. See also 9.1.3.

24. These officials are:

     (a)   technical officials of the Public Hygiene Administration with regard to the health protection of the population;

     (b)   engineers from the Safety at Work Administration and Engineers from the Mines Administration with regard to the safety of the workers in the vicinity of nuclear installations;

     (c)   doctors of the Administration of Hygiene and Medicine at Work with regard to the medical surveillance of workers and working conditions in nuclear installations;

     (d)   doctors of the Administration of Hygiene and Medicine at Work with regard to the medical surveillance of workers and working conditions in nuclear installations;

     (e)   the responsible officials of the Belgian National Railway Company with regard to transportation carried out by the railway company;

     (f)   the officials and agents of the Transport Administration invested with a judicial police mandate, with regard to transport;

     (g)   engineers and technical officers of the Administration for Highways and Bridges in charge of the navigation service, the maritime commissioners and their agents, the officials and inspectors of the Maritime Inspection Service, Harbour Masters and captains with regard to water transport;

     (h)   postal officials appointed by royal or ministerial nomination with regard to the postage of radioactive matter;

     (i)   officials of the Aeronautical Administration attached to the airports with regard to air transport;

     (j)   officials and agents of the Customs Administration with regard to the importation, transit and transport within ten kilometers of the frontier.

25. *Moniteur belge*, 25 August 1981.

26. Council of State, 27 June 1980, no. 20.485, *Burgomaster of the Town of Huy* v. *The Belgian State.*

27. Report to the King, Royal Decree of 30 March 1981 defining the task and establishing

the working rules of the Public Company for the Management of Radioactive Wastes and Fissile Materials, *Moniteur belge*, 5 May 1981.

28. N*Moniteur belge*, 23 August 1966.
29. *Moniteur belge*, 20 November 1975.
30. *Moniteur belge*, 17 October 1978.
31. See the Ministerial Decree of 24 April 1964, *Moniteur belge*, 22 May 1964.

10
Product Control

10.1 THE LAW OF 24 JANUARY 1977 RELATING TO THE PROTECTION OF THE HEALTH OF CONSUMERS AS REGARDS FOODSTUFFS AND OTHER PRODUCTS

10.1.1 General

10.1.1.1

The basic law on the control of products is the Law of 24 January 1977 relating to the health protection of consumers with regard to foodstuffs and other products,[1] replacing the Law of 20 June 1964.

In the application of this law:

(1) foodstuffs are all products or substances intended for human consumption, including tonics, salt and condiments, as well as natural aromatic products and their constituents and synthetic aromatic products with an identical chemical formula;
(2) other products are:
     (a) additives;

(b) substances and objects intended to come into contact with foodstuffs;
(c) detergents and cleaning and maintenance products which, in normal use, are likely to come into contact with foods;
(d) tobacco, tobacco-based and similar products;
(e) cosmetics;
(f) products which, when in use, could create a physiological effect either by the absorption of certain of their constituent parts, or by contact with the human body;
(g) aerosols and gas propellants used for foods and other products covered by (a) to (f).

10.1.1.2

First, this law authorises the King to control or prohibit the manufacture, exportation and marketing of *foodstuffs*. This power involves the possibility of determining the composition of foodstuffs, specifying the corresponding denominations and controlling the labelling and information required, on the suggestion of the Minister of Public Health.

Second, on the suggestion or opinion of the Superior Council of Hygiene, the King can control and prohibit the marketing of dietary foods, vitamins, oligo-elements and other foods.

Third, the King can designate certain dietary foods to be subject to registration under conditions and according to rules determined by him.

10.1.1.3

The law also enables the King to control four potential external *sources of food contamination:*[2]

(a) individuals in contact with foodstuffs;
(b) objects and materials in contact with foodstuffs;
(c) premises where foodstuffs are manufactured or held;
(d) detergents and cleaning and maintenance products.

10.1.1.4

With regard to *additives,* the King is authorised to establish a list of authorised additives which specifies the foods in which, and the proportions in which the additive is authorised. He also fixes purity criteria for additives and the method of expressing maximum content. Mention of the inclusion of the additive and the quantity involved can be required on the packaging of the food.

10.1.1.5

On the suggestion or opinion of the Superior Council of Hygiene, the King can control, prohibit or restrict the presence of *contaminants* in foodstuffs.

The marketing of foodstuffs containing prohibited contaminants or higher quantities of contaminants than those authorised by the King is forbidden.

> It is useful to note that the method of control with regard to contaminants is different to that contained in article 4 for additives. If the same strict rules were applied to contaminants as are applied to additives, i.e. prohibition on marketing any foods containing a contaminant without prior authorisation, they would, if strictly implemented, result in the blocking of production and marketing of foodstuffs.[3]

10.1.1.6

Cosmetics, tobacco, tobacco-based and similar products, as well as *aerosols* and gas propellants can also be controlled at the manufacturing and marketing stage; the hygiene of individuals and premises, and the composition of packaging and containers can also be controlled.

The King can control the marketing and exportation but not the manufacture of *'usual products'*—products which, through their use, could exercise a physiological effect either by absorption of certain of their constituent parts, such as toys, colouring crayons, etc., or by contact with the human body, such as certain textiles. For all these products the King can determine the substances which they must not contain or must only contain in a quantity determined by him, as well as the limits and conditions to which the presence of these substances are subject.

The law also envisages the possibility of a registration system for certain cosmetics, under conditions to be fixed by the King.

10.1.1.7

The King can control all forms of *advertising* concerning the health and dietary aspects of foodstuffs, the health properties of cosmetics, additives, detergents, and maintenance products; see the Royal Decree of 17 April 1980 concerning the advertising of foodstuffs.[4]

The King can also control and prohibit tobacco advertising; see the Royal Decree of 5 March 1980, amended by the Royal Decree of 22 September 1980.[5]

10.1.1.8

With regard to *labelling,* the law states that indications must be written in the language of the region where the products are sold *and* in one or several languages commonly used in Belgium. The procedure has been chosen in preference to a trilingual solution since certain products have only a limited sales area.

10.1.1.9 MONITORING THE APPLICATION OF THE LAW

The implementation of the provisions of this law and its implementing decrees is monitored by the burgomaster or his delegate and officials and agents specially appointed for this purpose by the King,[6] without impairing the powers of officers of the judicial police.

They can enter all premises and depots involved in the marketing of foodstuffs or other products covered by this law. When premises normally accessible to consumers are involved, this power is limited to the hours during which the premises are open to consumers.

They can enter at any time premises used for the manufacture of foodstuffs or other products covered by this law as well as premises in which such products are stored.

They can require the production of all commercial records and documents relating to foodstuffs and other products covered by the law, as well as all documents required under the decrees issued in implementation of the law.

They verify infringements of laws and decrees on the subject by means of written reports which are taken as evidence in the absence of proof to the contrary. A copy of the report is sent to the offender within ten days of the verification of the infringement.

Methods and conditions for taking samples, as well as methods of analysis, are determined by royal decree. Analysis costs are charged to the offender if the analysis reveals that the law is being infringed.

10.1.1.10

The maximum *penal sanctions* provided extend to imprisonment for a year and/or a fine of 15,000 francs.

Administrative fines are also applied, on payment of which action by the public authorities is cancelled; such fines are not entered in the record of convictions.

10.1.1.11

Article 18 of the law controls in detail the destruction of foodstuffs and other spoiled products which are harmful or declared to be harmful, as well as their removal, sequestration and/or seizure, and similarly the control of foodstuffs and other products presented for importation.

10.1.1.12

By means of decrees debated in the Council of Ministers, the King can issue all necessary measures in implementation of the law to ensure the fulfilment of obligations resulting from international treaties and acts issued as a result of such treaties; these measures can include the repeal or amendement of legal provisions.

10.1.1.13

The provisions of the law do not prejudice the rights granted to the communal authorities by the legislation in force to ensure the quality of foodstuffs and their cleanliness, or the legislation aimed at repressing infringements of the regulations issued by these authorities.

10.1.2 Additives

10.1.2.1

The Royal Decree of 27 July 1978[7] fixes the list of authorised additives in foodstuffs. The only additives—in categories other than artificial aromatic substances—authorised for foodstuffs are those mentioned in the annex of this decree, under the following conditions:

(1) The maximum content mentioned in column I must not be exceeded.

(2) The use of each additive is limited to the foodstuffs mentioned in column II of the annex.
(3) The use of additives is linked to the possible restricting conditions in column III of the annex.
(4) The additives must satisfy the purity criteria which may be fixed in other regulatory provisions. The presence of an additive is permitted in a prepared or made up food containing as an ingredient the food for which this additive is explicitly authorised as a result of a provision in the annex.[8] This transferred additive is authorised in the prepared or made up food up to a maximum content equal to the product of the content of the ingredient in the made up or prepared food and the content of the authorised additive in this ingredient.

Article 3 of the royal decree controls in detail the simultaneous use of additives, while article 4 deals with the method of expressing contents which varies according to whether they are chemical substances and their salts, phosphates or other additives.

10.1.2.2

The marketing and labelling of additives are controlled by the Royal Decree of 2 October 1980.[9]

It is forbidden to market additives, mixtures of additives or mixtures of additives with solvents or supports:

(1) if the additives, solvents and supports are not included in annexes I and III of this decree.
(2) if they are not prepacked;
(3) if they contain pathogenic micro-organisms or toxins of a microbial origin;
(4) if the additives do not satisfy the purity criteria envisaged in annex II of the decree;
(5) if products other than the solvents and supports permitted for each additive in annex III are included in the additive mixture;
(6) if the solvents and supports contained in the mixture mentioned under (5) do not satisfy the requirements contained in the regulations concerning these products.

It is also forbidden to market additives, mixtures of additives or mixtures of additives with solvents or supports if they do not carry the information indicated in article 4 of the decree.

The decree also contains the following prohibitions:

(1) It is prohibited to use on or in the proximity of products covered

187

by this decree, or on commercial documents relating to such products, names, indications, displays, signs or any other form of presentation which is not true concerning the nature, composition or characteristics of these products or their properties on the question of hygiene or public health.

(2) It is forbidden to make recommendations on or in the proximity of products covered by this decree, or in commercial documents concerning the use of these products in foodstuffs where their use is forbidden as a result of the regulation specifying the list of authorised additives for foodstuffs.

(3) It is forbidden for products which are also used as vitamins or nutrients to make use or allusion of any kind to these properties when the products are marketed as additives.

10.1.3 Aerosols

The Royal Decree of 14 April 1978[10] fixes the requirements with which aerosols must comply.[11]

It is forbidden to market aerosols which:

(1) do not satisfy the requirements laid down in the annex of this royal decree;[12]

(2) do not carry the indications required by article 3 of the decree;

(3) carry marks or inscriptions which could cause confusion with the sign '3' (reversed epsilon) permitted only for aerosols meeting strict requirements);

(4) do not carry an indication of the net content in weight and volume.

The Royal Decree of 28 March 1982 implementing Council Decision 80/372/EEC of 26 March 1980, concerning chlorofluorocarbons in the environment, requires manufacturers of aerosols to reduce, from 1 January 1982, the use of chlorofluorocarbons F11 and F12 in the filling of aerosol cans by at least 30% compared with the 1976 level. The industry can no longer increase its production capacity for chlorofluorocarbons F11 and F12 (12 *bis*).

10.1.4 Cosmetic products

10.1.4.1

As a result of the Royal Decree of 10 May 1978 relating to cosmetic products,[13] it is forbidden to market cosmetic products:[14]

(1) containing one or more of the substances listed in annex I of the decree;
(2) containing one or more of the substances listed in annexes II and III of the decree in excess of the limits or outside the conditions indicated in these annexes;
(3) containing one or more colours other than those listed in annexes IV and V if these products are to be applied in the proximity of the eyes, on the lips, inside the mouth or on the external genital organs;
(4) containing one or more of the colours listed in annexes IV and V in excess of the limits or outside the conditions indicated in these annexes, if these products are to be applied in the proximity of the eyes, inside the mouth or on the external genital organs;
(5) liable to present a danger for the health of the consumer when applied under normal conditions of use.

It is also forbidden to market cosmetic products which do not carry the indications required by article 3 of the decree. Finally, it is forbidden to use names, indications, displays, signs or any other form of presentation on the labelling of cosmetic products, in their vicinity, on commercial documents, prospectuses or any other form of advertising relating to these products which could result in an error concerning the nature, composition, method of manufacture or characteristics of the products or which could attribute to them remedial or prophylactic properties or characteristics relating to health that these products do not possess.

10.1.4.2

A Royal Decree of 17 August 1977[15] prohibits the marketing of any product for oral use intended exclusively or primarily to alter the colour of the skin.

10.1.5 Detergents

As a result of the Royal Decree of 23 March 1977 relating to the degree of biodegradability of certain surface active agents in detergents,[16] it is forbidden to import, market, or use detergents if the average biodegradability of the surface active agents they contain is less than 90% for each of the following categories: anionic, cationic, non-ionic and ampholytic. The use of surface active agents with an average biodegradability rate of at least 90% should not, under normal conditions or use, result in harm to the health of man, animals or plants.

The permitted tolerance with regard to the biodegradability rate of anionic surface active agents is fixed at 10%.

10.1.6 Labelling of prepacked foodstuffs

The basic rule on this subject is contained in article 2 of the Royal Decree of 2 October 1980 relating to the labelling of prepacked foodstuffs:[17]

> Without prejudice to the special provisions relating to certain foodstuffs, it is forbidden to market prepacked foodstuffs which do not carry the following information, under the conditions and apart from the exemptions envisaged in articles 3–10:
> (1) the selling name of the product:
> (2) the list of ingredients;
> (3) the minimum durability date;
> (4) special conditions for storage and use;
> (5) the name and address of the manufacturer or processor, or of the trader established in one of the Member States of the European Community;
> (6) method of use when its omission could result in an unsuitable use of the food;
> (7) the place of origin or production when its omission would be liable to mislead the consumer concerning the real origin or place of production of the food;
> (8) the net quantity.

10.2 THE LAW OF 11 JULY 1969 RELATING TO PESTICIDES AND RAW MATERIALS FOR AGRICULTURE, HORTICULTURE, FORESTRY CULTIVATION AND LIVESTOCK BREEDING[18]

10.2.1 General

10.2.1.1

This law has two objectives:

(a) to safeguard the interests of manufacturers, producers, purchasers,

preparers, livestock breeders, distributors, users and consumers by measures aimed at preventing fraud and adulteration, at repressing processes which distort the normal conditions of commercial competition and with a view to encouraging, improving and protecting crop and livestock production;
(b)   to safeguard public health.

10.2.1.2

The discussion here is limited to an explanation of the provisions relating to the public health interests. These provisions concern *pesticides,* that is, 'products intended to ensure the destruction or prevention of the harmful action of animals, plants, micro-organisms or viruses'.

10.2.1.3

The King is empowered to take all measures he considers to be necessary or useful.

He can:

(1)   determine the conditions of production, manufacture, processing, preparation, composition, particularly of active substances, storage, transport, quality, effectiveness, conditioning and advertising which must be satisfied by pesticides, as well as the qualities they must possess if they are to be marketed, purchased, offered for sale, displayed, sold, stored, used, prepared, transported, offered as free or conditional gift, imported, exported or permitted in transit;
(2)   determine the conditions of use for pesticides regarding the harmful consequences which could result for human health;
(3)   fix the maximum quantities for residues of active substances which can be left by pesticides and their products after eventual degradation;
(4)   subject the activities of individuals carrying out the operations indicated under (1) to prior authorisation or approval granted by the Minister of Public Health or by an organisation or official delegated by him for this purpose;
(5)   fix the conditions to be satisfied by individuals who are subject to control under this heading, in order to obtain and keep the authorisation or approval provided for under (4);
(6)   determine the imprints, leaded seals, seals, labels, certificates, attestations, documents, signs, packaging, names or other indications or documents establishing or attesting that the conditions under (1) are fulfilled; etc.

10.2.1.4

With regard to monitoring of the application of this law, its sanctions, the legal provisions—articles 6–13—are similar to those of the law of 24 January 1977, with certain peculiarities:

(a)   In addition to the penal sanctions, the law provides for *administrative sanctions,* particularly the refusal or withdrawal of authorisation approval from the offender or the legal person of whom the offender is an employee or agent.

(b)   *Provisional seizure:* the agents appointed in article 6 of the law can, as an administrative measure and for a period fixed by the King, provisionally seize a pesticide which they presume does not conform with the provisions of a decree issued in implementation of the law, for the purpose of having it analysed. This seizure is cancelled on the order of the agent who has carried out the seizure of the pesticide, or alternatively when the period of seizure expires.

(c)   *Definitive seizure:* in case of infringement, pesticides can be seized by the agents appointed in article 6 of the law. If the seized pesticides are perishable, they can be sold or returned to the owner on payment of an indemnity, provided such an action is permitted by the public health requirements; in this case they can only be disposed of in accordance with the instructions given by the agents designated by the minister responsible. The sum obtained is deposited with the court until a judgement has been passed concerning the infringement. This sum represents the seized pesticide for the purposes of its confiscation and eventual restitution to the party involved.

(d)   *Confiscation:* in the event of an infringement being confirmed, the court can order the confiscation and destruction of the seized products. Confiscation and destruction are always ordered if required by the nature or composition of the product. The destruction of seized products ordered by the court is carried out at the expense of the offender.

(e)   *Publication of the judgement:* the court can also order the publication of the judgement in one or more daily newspapers and the erection of posters in locations and for a period determined by the court, announcing the judgement, at the expense of the offender.

10.2.2 Pesticides

10.2.2.1

The Royal Decree of 9 June 1975 relating to the storage, marketing and use of pesticides and agricultural chemicals, amended by the Royal Decrees of 22 October 1976 and 23 March 1977[19] institutes a procedure of control by approval—by the Minister responsible for Agriculture after consultation with an approval committee and with the agreement of the Minister of Public Health—for agricultural chemicals, including pesticides for agricultural use, and a procedure of control through authorisation by the Minister of Public Health, with the agreement of the Superior Council of Public Hygiene for non-agricultural pesticides.

This approval or authorisation is individual and non-transferable; it is generally granted for a maximum period of ten years.

It is prohibited to use an approved or authorised product for purposes or under conditions other than those imposed by the minister concerned at the time of approval or authorisation.

10.2.2.2

With regard to toxic products, only individuals who satisfy the conditions fixed by the decree—the possession of certain diplomas or proof of the necessary qualifications, access to suitable premises, materials and equipment—can be approved as salesmen or users of the products contained in lists A, B and C of annex II of the decree.

10.2.2.3

With regard to *labelling,* the packaging of agricultural chemicals and pesticides for non-agricultural use must carry the fifteen indications listed in article 18 of the Royal Decree of 9 June 1975. The *packaging* must be designed and constructed in such a way as to:

(1) prevent any leakage of contents, with the exception of that arising from the functioning of a regulatory safety device;
(2) avoid any damage from the contents, or any formation of dangerous combinations with the latter;
(3) stand up to handling and transport requirements.

10.2.2.4

Finally, the Royal Decree of 9 June 1975 also contains provisions relating to measures for the protection of workers and measures regarding sampling[20] and the period for seizure by the administrative authorities (three months).

10.2.3 Fertilizers

10.2.3.1

The marketing of manure and fertilizers is controlled by the Royal Decree of 6 October 1977[21] which is based not only on the law of 11 July 1969 but also on recommendation M (68) 12 of the Committee of Ministers of the Benelux Economic Union of 29 January 1968 concerning the control of exchanges of manure, lime, organic fertilizers and associated products within the Benelux area, modified by recommendation M (69) 17 of the Committee of Ministers of the Benelux Economic Union of 14 April 1969 and also on Directive 116/76/EEC of 18 December 1975 concerning the harmonisation of legislation on fertilizers in Member States.

This decree prohibits the marketing of manure and fertilizers, as well as all products with a specific action to stimulate crop production, if they are not included in the table annexed to the decree.

The products contained in this table can only be marketed under the denomination of the type specified in column (a). In addition, they must conform to the description given in column (b), to the criteria and requirements in column (c) and possess the substantial qualities listed in column (d), the contents of which must be guaranteed.

It is prohibited to use the name 'EEC fertilizer' for products which do not conform to the criteria and requirements fixed for such fertilizers in the decree and in chapter I, division I of its annex.

10.7.3.2

The decree regulates in detail the guarantees—information and indications, documents and transport, closed packaging fitted with a seal etc., the method of taking and analysing samples,[22] the peanl sanctions and administrative measures (seizure for a maximum period of thirty days).

10.2.4 Substances to be used for animal feed

In its annex, known as 'codex', the Royal Decree of 12 July 1972[23] contains a table of substances for use as animal feed.

It prohibits the marketing of substances for use as animal feed unless they are included in column (a) of chapters I, II and III of the codex.

The substances listed in the codex can only be marketed under the name contained in column (a). In addition, they must correspond with the definitions given in column (b), the standards specified in column (c) and satisfy the special requirements of the codex.

Anyone importing, manufacturing or preparing the substances covered by chapters II and III of the codex must have the prior approval of the Minister of Agriculture.

10.3 OTHER LEGAL TEXTS RELATING TO THE PRODUCTION AND MARKETING OF DANGEROUS SUBSTANCES

10.3.1

The specific legislation already described is usefully extended by provisions of a more general character, such as:

the controls relating to dangerous, dirty and noxious establishments (General Regulation for Protection at Work, heading I); see Chapter 2;

the General Regulation for the Protection of the Population and Workers against the Danger of Ionising Radiation (Royal Decree of 28 February 1963), based on the outline Law of 29 March 1958; see Chapter 9;

the General Regulation on the Manufacture, Storage, Transport and Use of Explosive Products (Royal Decree of 23 September 1958), based on the Law of 28 May 1956 relating to substances or mixtures which are explosive or liable to ignite, and to the machines containing them.

Broadly speaking, this last regulation greatly resembles the General Regulation on Classified Establishments: it establishes an obligation to obtain authorisation which can be withdrawn at any time.

It should be pointed out here that only explosive substances and mixtures which have been recognised by the Minister of Economic Affairs can be manufactured in Belgium.

10.3.2

Heading III, chapter II of the General Regulation for Protection at Work contains specific measures in section II, which apply to chemical industries:

certain toxic substances—39 in total—can only be supplied in packages which mention precisely, in a very visible and legible manner, that they contain such substances (article 393);

it is forbidden to manufacture, sell and store matches containing white phosphorous (article 395);

in certain industries the use of paints containing lead can be prohibited or subject to restrictions (article 397);

manufacturers of carbonate of lead, lead sulphate and other white lead pigments must keep a register in which they record the monthly quantity manufactured, as well as the sales per product and the names of the purchasers; these substances can only be released to individuals who hold a purchasing authorisation (article 398–400).[24]

10.3.3

As a result of the Law of 24 February 1921, the government is authorised, in the interests of hygiene and public health, to control and monitor the import, export, manufacture, transport, storage, sale and offer for sale, delivery and conditional purchase of toxic, soporific, narcotic, disinfectant or antiseptic substances as well as the cultivation of plants from which these substances can be extracted.[25]

The following are among the implementing decrees:

Royal Decree of 30 December 1930 concerning the traffic of soporific and narcotic substances (prior authorisation);

Regent's Decree of 6 February 1946 controlling the storage and output of poisonous and toxic substances;

Royal Decree of 6 June 1960 relating to the manufacture, preparation, wholesale distribution and dispensing of medicines (prior authorisation);

Royal Decree of 12 April 1974 relating to certain operations concerning substances with a hormonal, anti-hormonal or antibiotic action (prior authorisation);

Royal Decree of 31 May 1976 concerning certain psychotropes (prior authorisation).

10.3.4

The Royal Decree of 24 May 1982 concerning the placing on the market of substances which could constitute a hazard for man or the environment[26] implements EEC Directives, in particular Directive 79/831/EEC of 18 September 1979.

Before placing a dangerous substance on the market, any manufacturer or importer is required to submit to the Minister of Public Health a notification including:

a technical dossier supplying the information necessary for evaluating foreseeable risks, whether immediate or delayed, which the substance may entail for man and the environment, and containing the information and results of studies referred to in annex VII of the decree, together with a detailed and full description of the studies conducted and of the methods used or a bibliographic reference to them;

a declaration concerning the unfavourable effects of the substance in terms of the various uses envisaged;

the proposed classification and labelling of the substance in accordance with the provisions of the Royal Decree of 9 April 1980 and 19 March 1981 relating to the classification, packaging and labelling of dangerous substances and preparations, and with those of the Royal Decree of 24 May 1982;[27]

proposals for any recommended precautions relating to the safe use of the substance;

a summary of the notification.

A 'Committee on Dangerous Substances' composed of officials of different ministerial departments is appointed and attached to the Ministry of Public Health. This Committee is responsible for examining the notification; it advises the minister on the completeness of the application. The minister takes the final decision on notification within forty-five days of receiving the application.

The dangerous substance cannot be placed on the market during these forty-five days, or while the notification is considered as incomplete or not in accordance with the procedure laid down in the Royal Decree of 24 May 1982.

10.3.5

The *transport* by road of dangerous merchandise—with the exception of explosive and radioactive matter—is controlled by provisions in the Royal Decree of 15 March 1976, amended by the Royal Decrees of 1 June 1977, 7 April 1978 and 25 September 1978.[28]

The basic principle is that the provisions of the European Agreement relating to the International Transport by Road of Dangerous Merchandise (ADR), signed in Geneva on 30 September 1957 and approved by the Law of 10 August 1960, are applicable to all road transport of merchandise, even if the transport is not international under the meaning of this Agreement.

The annexes of the European Agreement, which are regularly amended and amplified, contain general provisions (transport authorisation, control over the construction of vehicles etc.) and specific provisions relating to certain types of dangerous merchandise.

10.3.6

The Law of 12 April 1965 controls the transport of gaseous and other products by pipeline.[29]

Certain provisions of this law have been extended to the transport by pipeline of the following materials:

industrial waste water along the Albert canal (Royal Decree of 19 February 1971, *Moniteur belge*, 20 July 1971);

liquid hydrocarbons and/or liquefied hydrocarbons (Royal Decree of 15 June 1967, *Moniteur belge*, 22 June 1967);

gaseous oxygen (Royal Decree of 14 March 1969, *Moniteur belge*, 1969);

brine, caustic soda and liquid residues (Royal Decree of 15 June 1967, *Moniteur belge*, 22 June 1967).

Notes

1. *Moniteur belge*, 8 April 1977.
2. See the Royal Decree of 13 November 1978 relating to hygiene during manufacture and marketing of foodstuffs, *Moniteur belge*, 7 December 1978.
3. Draft law, Explanation of reasons, commentary on article 5, *Doc. parl.*, Chamber of Representatives, 1974–75, doc. no. 563/1.
4. *Moniteur belge*, 6 May 1980.
5. *Moniteur belge*, 14 March 1980 and 1 October 1980.
6. The Royal Decree of 1 December 1977 makes food inspection officials and food and meat control agents responsible for monitoring the Law of 24 January 1977 and its implementing decrees; *Moniteur belge*, 7 February 1978.
7. *Moniteur belge*, 20 October 1978; this royal decree was amended by the Royal Decrees of 17 April 1980 and 2 October 1980, *Moniteur belge*, 29 July and 11 October 1980.
8. This transfer of additives is, however, not authorised for foodstuffs for feeding babies under three months old.
9. *Moniteur belge*, 16 January 1981. The following are not covered by this decree:
   (1) vitamins when they are used for their vitamin properties;
   (2) nutrients when they are used for their nutritive properties;
   (3) technological auxiliaries;
   (4) artificial aromatic substances;
   (5) artificial sweeteners marketed only for direct delivery to the consumer.
10. *Moniteur belge*, 25 May 1978. This royal decree was issued in implementation of the Law of 24 January 1977 and Directive 75/324/EEC of the Council of European Communities of 20 May 1975.
11. 'Aerosols' are: non-reusable containers in metal, glass or plastic containing compressed, liquefied or dissolved pressurised gas, with or without liquid, paste or powder, and with a device enabling the contents in the form of solid particles or liquids in suspension to be released in the form of a gas, foam, paste, powder or liquid.
12. There are several exemptions of little importance.
12(a).*Moniteur belge*, 24 April 1982.
13. *Moniteur belge*, 1 September 1978. See Directive 76/768/EEC of 27 July 1976.
14. 'Cosmetic products' are: all substances or preparations intended to come in contact with the various surface areas of the human body, the skin, hair, body hair, nails, lips and external genital organs or with the teeth and oral mucous areas with the sole or principal purpose of cleaning, perfuming, protecting in order to keep them in good condition, altering the appearance or correcting bodily odours. Annex I of the royal decree contains a 'negative list' of 357 substances; annex II contains a list of 29 substances; annex III a list of 24 substances; annex IV a list of 94 substances and annex V a list of 38 substances.
15. *Moniteur belge*, 16 September 1977.
16. *Moniteur belge*, 9 June 1977. This royal decree was issued in implementation of the Law of 28 February 1970 in approval of the European Agreement on the limitation of the use of certain detergents in washing and cleaning products, made in Strasbourg on 16 September 1968, on the one hand, and the Directives of 22 November 1973 of the Council of the European Communities, nos. 73/404/EEC and 73/405/EEC on the other hand.
17. *Moniteur belge*, 11 October 1980.
18. *Moniteur belge*, 17 July 1967; err. 29 July 1967. See also the Ministerial Decree of 17 March 1980 appointing the officials responsible for the monitoring of the implementation of the Law of 11 July 1969, *Moniteur belge*, 5 April 1980.
19. *Moniteur belge*, 4 November 1975, 28 October 1976 and 6 May 1977. See also

Directive 78/631/EEC as amended by Directive 81/187/EEC, *OJ* L206, 29 July 1978 and *OJ* L77, 2 April 1981.

20. See also the Royal Decree of 8 February 1980 relating to the approval of laboratories for the analysis of pesticides for non-agricultural use, *Moniteur belge*, 28 June 1980.

21. *Moniteur belge*, 30 December 1977; amended by the Royal Decree of 18 September 1978, *Moniteur belge*, 24 October 1978.

22. See also the Ministerial Decree of 7 October 1977, *Moniteur belge*, 30 December 1977.

23. *Moniteur belge*, 19 January 1973. This royal decree has been amended several times; see the Royal Decrees of 4 March 1976, 4 November 1976, 21 December 1977 and 5 February 1979, *Moniteur belge*, 24 July 1976, 12 February 1977, 17 January 1978 and 28 February 1978. See also the Ministerial Decree of 3 May 1978 relating to the marketing and use of substances intended for animal feed, *Moniteur belge*, 30 September 1978, amended by Ministerial Decree of 26 October 1979, *Moniteur belge*, 26 January 1980; err. 6 May 1980.

24. The same applies to the purchase, sale and use of cyanogen, hydrocyanic acid and its salts (article 696–698).

25. *Moniteur belge*, 6 March 1921, amended by the Laws of 22 July 1974, 9 July 1975 and 1 July 1976.

26. *Moniteur belge*, 2 July 1982.

27. The following substances and preparations are considered to be dangerous: explosive, oxidising, flammable, toxic, harmful, corrosive, irritant, carcinogenic, teratogenic, or mutagenic substances and preparations, as well as those which are dangerous to the environment. The provisions of the Royal Decree of 24 May 1982 do *not* cover:
    (a) substances to be used exclusively as foodstuffs;
    (b) substances to be used exclusively as animal feed;
    (c) substances to be used exclusively as medical products and narcotics;
    (d) substances to be used exclusively as phytopharmaceutical products;
    (e) substances in the form of wastes;
    (f) substances in transit which are under customs supervision, provided they do not undergo any treatment or processing;
    (g) the carriage of dangerous substances by rail, road, inland waterway, sea or air;
    (h) radioactive substances.
    The list of dangerous substances is identical to the list in Chapter 3 of Title III of the General Regulation for the Protection of Labour (Royal Decrees of 29 May 1978 and 13 February 1981).

28. *Moniteur belge*, 12 June 1976, 10 June 1977, 18 April 1978, and 3 October 1978.

29. *Moniteur belge*, 7 May 1965.

11

Environmental Impact Assessment

11.1

The procedure for assessing impact on the environment is based on a description, together with an indication of the alteration, of the consequences for the environment of a programme, project or initiative which a public authority or private individual proposes to carry out. This 'cost-benefit' analysis is submitted to an inquiry and to evaluation by independent experts.

To date, Belgian law has not recognised the obligation for an impact assessment. Nevertheless, as already indicated, the underlying philosophy of the environmental impact study is not unknown to Belgian jurists because it has been in use in Belgium since 1810—to a very modest degree, it is true—for the control of dangerous, dirty or noxious establishments.

As is shown clearly in a document from the association 'Interenvironment—Bond Beter Leefmilieu', environmental impact studies are intended to improve upon present public inquiries by acting as:

a *planning* instrument for the person in charge of the project because he must take account of the environmental consequences of the project during the planning phase. The impact study must indicate how the best possible solutions for the prevention of harmful effects have been arrived at.

an *information* instrument by means of the publicity given to the case in the context of the public inquiry and by the holding of public audiences which, if necessary, can enable alternative solutions to be put forward.

a *decision-making* instrument by way of the information provided and the opinions gathered by the impact study, together with, as a corollary, a rationalisation and co-ordination of the administrative procedures already in existence.

a *control* instrument through the laying down, from the information collected by the impact study, of a procedure of periodic monitoring to ensure that the work is being carried out in accordance with the principles of the project.

11.2

The previous Minister of Public Health and the Environment repeatedly announced his intention to present a draft law to parliament. However, the fall of the government prevented him from doing so. The present minister has not yet made known his intentions on the subject.

All we can therefore do is to indicate the broad outlines of the preliminary draft. This draft was largely inspired by the draft proposal for a directive of the Council of the European Communities (COM (80) 313 FINAL).

11.3

The preliminary draft restricts its field of application to plans which are already subject to obligatory authorisation, without making any distinction with regard to the initiator who can be either a public administration or a private individual. The obligation for a prior evaluation of the possible impact on the environment of proposed activities does not therefore include development plans and programmes.

11.4

Projects liable to have a major impact on the environment as a result of their nature, size or location are subject to an appropriate evaluation of that impact.

The proposed draft law enumerates three categories of project and two types of impact evaluation: the 'impact study' and the 'notice of impact'.

A 'notice of impact' is sufficient for projects which are not likely to have a major impact on the environment—List A. Its content will be established by royal decree.

An 'impact study' is necessary for other projects and must be carried out by an approved institution, under the auspices of the head of the project. This study should contain:

(1) a description of the proposed project and of the original state of the environment which runs the risk of being damaged by the project;
(2) the foreseeable consequences for the environment;
(3) a description of the possible alternatives and the reasons for the choice of that particular project;
(4) a description of the measures aimed at preventing, restricting or compensating for the effects on the environment;
(5) a non-technical summary.

11.5 PROCEDURE

The authorisation application, the notice of impact or impact study, the various opinions obtained and, should the occasion arise, any additional useful information, are all made available to the public. In principle the public inquiry lasts for thirty days, but this is reduced to fifteen days for projects on List A.

In the light of the data received from the head of the project and the information and comments received during the course of the various consultations, the minister responsible evaluates the extent of the probable impact of the proposed project on the environment. He then indicates the measures that will be necessary to limit or prevent harmful effects on the environment.

Any interested party can make an administrative appeal to the minister against the form and content of the impact report established by the minister. This appeal has a suspensive effect and the minister must take a decision within forty days.

11.6

The procedure for evaluating the impact of the environment is not part of the existing procedures for authorisation applications;[1] it is superimposed on them.

Once the impact report has been established, the relevent authorities must take a decision on the authorisation application. In all cases, this decision must be justified, regardless of whether it is positive or negative, and made public.

Any authorisation which is granted contrary to the legal provisions relating to the evaluation of environmental impact is null and void.

11.7

It should be noted that a negative environmental impact evaluation does not necessarily lead to a negative decision on the part of the authorities granting the authorisation. The objective of the impact assessment is to evaluate a project according to ecological criteria, that is, a complete, clear knowledge, discussed with an open mind, of the repercussions which the project could have on the environment. At the time of the final authorisation decision, other factors, such as economic considerations—e.g. regional development—or social considerations, can intervene and even outweigh the importance of the impact assessment.

Note

1. See M. Van Holder, *De onderneming en haar milieuvergunningen,* in Leefmilieu, 1978, no. 5, pp. 149–155.

12

Compensation for Damage to the Environment*

Compensation for damage caused by disturbance of the environment is the subject of this chapter. The central themes are the liability for personal fault (article 1382 Civil Code), liability for defects in objects for which a person is responsible (article 1384(2)) and the theory of neighbourhood disturbance, as well as certain strict liabilities introduced by the legislator for specific damage.

This chapter restricts itself to the problems relating to Belgian internal law. Transfrontier pollution and the pollution of international waters are not examined here, nor is the specific problem of nuclear responsibility.

12.1 SUBJECTIVE LIABILITY

In Belgian law, the responsibility for personal fault (article 1382) plays a major role in the question of compensation for damage to the environment. On this subject, the following factors are examined: (1) the fault, (2) the damage, (3) the causality link, (4) forms of compensation and (5) the responsibility of the public powers.

Of minor importance is the responsibility for damage caused by defects in objects for which individuals are responsible and this question is examined later in the chapter.

*By Hubert Bocken, University of Ghent. This chapter is based on a report written for the OECD Environment Directorate.

12.1.1 Liability for personal fault

12.1.1.1 FAULT

The vague idea of fault, as an act which would not have been committed by a reasonable man, the good family man,[1] has, in matters of environmental damage, been confirmed above all by reference to the following specific criteria:

(a) the violation of a legal or regulatory requirement;
(b) the general obligation for care and attention;
(c) the abuse of right;
(d) its practical extent is also influenced by the justifying causes taken into consideration.

12.1.1.1.1 Fault through the violation of legal or regulatory requirements

The violation of a legal or regulatory requirement in itself constitutes a fault (except where there is a justifying cause).[2] It provides a fairly easy point of departure for a compensation action. This form of fault is of great interest in the field of ecological damage because of the spectacular increase in the number of environmental protection provisions at municipal and provincial level, as well as at national level.

The provisions which make certain dangerous activities subject to authorisation are of great practical importance. A large number of compensation actions for ecological damage are based on the violation of operating permits required under the General Regulation for Protection of Labour. This is even more important where certain obligations are imposed by this permit, such as not to produce dust, not to disturb the neighbours, etc. As soon as damage envisaged by this type of provision occurs, it constitutes a violation of the permit, a penal infringement and a civil fault.[3]

While the violation of a legal or regulatory obligation constitutes a fault, the fact that such obligations are respected does not however guarantee freedom from responsibility.

The general obligation for care and attention is binding at all times and can impose more severe requirements than the specific legal or regulatory requirements.[4]

The violation of a penal law requirement, while being a civil fault like any other, does not result in less important consequences at the procedural level.

The application for compensation can be associated with the penal procedure by the prejudiced party coming in as a civil party. This often enables the victim to benefit from the efforts of the Public Ministry in identifying the guilty party and establishing his guilt. This can obviously result in a considerable alleviation in the difficulty of providing proof and of meeting the cost of an action.

The victim can also take his action for compensation for damage resulting from a penal infringement to the civil court. In the event of further penal proceedings being instituted for the same infringement, the civil judge can suspend the procedure until he has finally decided on the penal action. The penal decision on the existence of the infringement has the force of a judgement for the civil judge.

In addition, there is a considerable difference between a compensation action based on a penal infringement and that based on a purely civil fault. As a general rule, the former has a limitation of up to five years after the infringement (ten years in the case of a crime),[5] whereas the latter can, in principle, be introduced up to thirty years after the damage occurred.

12.1.1.1.2 The general obligation for care and attention

People who conduct themselves without due care and attention in respect of others or the possessions of others are committing an offence. At this level it is essential to make a comparison between the actions of the defendant and the normal conduct of the good family man, and on this point the jurisprudence is very casuistic.

In certain cases, it simply announces that reasonable and normal care and attention is required and that the precautions taken should be sufficient under normal conditions.[6]

Reference to the state of the art and available technology is frequently made. Individuals who undertake industrial or building work must in all cases take the precautions indicated by the state of the art and technology.[7] This approach leaves even wider discretionary powers open to the judge with regard to practical requirements to be imposed: should the most advanced technology be used, or only that which is economically viable, or should only the usual precautions be taken? The response given to this question is very varied; furthermore, the judges frequently do not explain the criteria underlying their decisions in this respect. The general impression, however, is that preference is given to the use of the usual technology. It would seem that decisions which require that all possible precautions from a technical point of view, or even from an economic point of view, are taken to avoid the deterioration of the environment are less numerous. Some notable judgements in this context have been pronounced in recent years.[8]

In a fairly high number of cases the application of the idea of fault seems to have been greatly influenced by the risk and social merits inherent in the damaging activity. The underlying interests, in these cases, have become predominant, rather than the factors external to the situation or the parties in question, such as the technology available. Thus, those who fail to take precautions to remedy a dangerous situation, or who do not react to a danger resulting from an accident or case of force majeure, are committing a fault. Similarly, there have been decisions imposing responsibility on the polluter for failure to inform the possible victims of the dangers they risk from his activities and to indicate to them the precautions which should be taken.[9]

Recently, failure to carry out the necessary research to avoid pollution caused by certain industrial activities[10] has been considered to be a fault. In a limited number of cases, the judge has based his decision on a comparison (often implicit) of the social advantages of a certain technology and of the risks it involves. In some cases, mostly old cases, the responsibility of the polluter has been based solely on the fact that the damage caused is judged to be something which should have been avoided in all cases.[11]

12.1.1.1.3 Abuse of right

The importance for the protection of the environment of the idea of an abuse of rights was considerable in the nineteenth century and at the beginning of the twentieth century. Since that time it has considerably diminished because of the changing concepts on the subject of the idea of subjective right itself, which was generally understood to be an (exclusive) prerogative provided by objective right. It is also agreed that the application of a certain, apparently inappropriate, technology does not normally constitute the exercising of a prerogative resulting, for example, from the right of ownership, but merely the right of freedom of a more or less controlled action.[12] It will not, therefore, in general be necessary to investigate whether polluting activities constitute an abuse of right giving rise to responsibility: the general idea of fault will be sufficient.

The criteria used by the jurisprudence to determine whether the exercising of right damaging others can involve responsibility are therefore mentioned here more to provide a complete picture than because of their practical importance in compensation for environmental damage.

The three traditional criteria by which the jurisprudence determines whether the use made of a subjective right gives rise to responsibility, i.e. the intention to injure, the absence of interest and the choice of a method of exercising one's right which is most harmful to others, do not correspond with the actual situations in which pollution is generally produced.

Of greater use, but still fairly theoretical, is the situation presented by the jurisprudence whereby right is abused if there is a disproportion between the damage resulting for others and the exercising of a right by the defendant, and the extent of the interest which this right represents to him.

The abuse of right interpreted as the exercising of a right which is contrary to the social interest is not often used in the context of actions on responsibility for ecological damage.[13]

According to recent practice, the responsibility for damage caused by the exercising of a right should be determined on the basis of the general obligation for care and attention already mentioned. This theory is not, however, in accordance with the decree of the Court of Cassation of 10 September 1971,[14] according to which an abuse of right can result from the unreasonable exercising of a right.

12.1.1.1.4 Justifying causes

Among the causes of justification invoked in the context of compensation actions for environmental damage, the idea of a state of necessity is worthy of comment. According to this theory, those who commit an apparently illicit act for the purpose of avoiding serious damage to themselves or to others, should not be considered to be responsible. In the event of damage by pollution, industrialists sometimes claim economic necessity as a justification. The cost of anti-pollution measures would be beyond their financial means; in other words, the pollution is necessary in order to safeguard the higher interests of the industry and the workers. This argument has not been favourably received by the jurisprudence. It was rejected in 1968 by the Police Court of Verviers and in 1976 by the Court of Gand.[15]

The same position had already been taken relating to municipalities responsible for polluting a stream by the discharge of untreated sewage.[16]

12.1.1.2 DAMAGE

12.1.1.2.1 General

In order to proceed with a responsibility action, the vicitm must have suffered damage, including damage to a legitimate right. Damage to a right does not therefore have to be proven.[17]

It is not necessary for the damage resulting from pollution to reach a certain intensity or to be abnormal. This requirement, which is justified in the context of objective responsibility, is hardly acceptable in the context of Article 1382.

A part of the jurisprudence[18] ignores these principles in application of article 1382 and requires the disturbance to exceed the normal level of neighbourly tolerance. This is criticised by the doctrine.[19]

It is, however, necessary for the damage, even future damage, to be certain. This places a number of forms of ecological damage (such as damage to the ozone layer by aerosols and its consequence for health) which, although disputed by ecologists, have not yet been established with certainty, outside the scope of actions for responsibility.

The requirement for the damage to be personal constitutes a substantial limitation on the effectiveness of responsibility actions with regard to compensation for ecological damage.

It is, of course, certain that harm to the environment constitutes a personal damage as soon as the victim suffers even minimal injury either to his physical well-being or his property.

Although the jurisprudence does not appear to have yet decided on this point, the view can also be defended that disturbance of the use (even when shared with others) that is made of natural assets (such as parks or river banks) can constitute personal damage.[21]

The situation is quite different with regard to attacks on natural assets for which the complainant cannot claim a personal use, such as the building property of others, parts of public property which are not accessible to the public, certain collective natural assets such as the oceans, the stratosphere and ecological concepts such as the ecological equilibrium of the environment. Activities which are merely contrary to the general interests or objectives (protection of nature, for example) of a specific person do not give rise to a responsibility action.[22]

12.1.1.2.2 Damage suffered by an association

The possibility for environmental protection associations to act for compensation for ecological damage is largely limited by the necessity for personal damage. Before the existence of a personal damage suffered by an association can be examined in depth, it is, of course, necessary that the action is judged to be acceptable.

In order for an association to be able to act through the civil courts it must, first of all, be of a legal character. To satisfy this requirement, use is generally made of the flexible legal form of association which is not formed for the purpose of making money. On the other hand, informal action groups are not allowed to act through the civil courts;[23] it is, however, possible for members of a group to give a mandate to one of their number to represent them in an action.[24] Nevertheless, at the level of the acceptability of action, the association must establish a personal interest before it can take legal action. It must demonstrate that the achievement of the objective of the action will be to its advantage. Even though decisions to the contrary are taken,[25] it is clear that an environmental protection association can invoke an interest in legally fighting against pollution: in fact, this action assists the association in the achievement of its objectives.[26]

The question of the existence of personal injury as a fundamental condition for a responsibility action is a completely different matter. There is no doubt that in such a case the same principles should apply as those involved in cases where physical damage is incurred.

There is therefore no problem if an association suffers damage to either its moral or material well-being. The same is true if it is disrupted in the course of its activities, a fact which justifies the jurisprudence in declaring numerous actions to be valid which are introduced by fishermen following pollution of their fishing waters.[27] As is the case when individuals are involved, an association cannot, however, claim as personal damage the fact that an activity is contrary to the interests or objectives which it pursues.[28] The same conclusion is valid for damage suffered by individual members of the association.[29] Unless the damage constitutes at the same time injury to the moral or material well-being of the association, it cannot be used to introduce an action for compensation.

The judges are not, however, unanimous in their application of these principles. At times there is confusion regarding the acceptability of, and the grounds for an action. In recent years, part of the jurisprudence has taken a more favourable line with regard to associations. Without explicit theoretical justification following certain infringements, certain judgements have declared the civil constitution of environmental protection associations to be well founded. Even though the aim of the association is in the first instance the repression of infringements and in spite of the fact that it is difficult to identify the injury that the association has suffered, it has been granted symbolic damage interests.[30]

Contrary to the situation in other countries, in Belgium the right of the associations to act at law has not, in recent years, been a subject which has preoccupied the legislator. Nevertheless, some ancient laws should be mentioned in this context:

A Law of 1898 gives professional associations the possibility of legal action to protect the interests of their members. The law enables fishing and hunting associations with the legal character of professional associations to act against certain damage to the environment which injures their professional interests.

The only law on the subject with a specific ecological aim is the Law of 12 August 1911 on the protection of the beauty of the countryside. It goes beyond the level of the association and enables every Belgian citizen to introduce civil actions for the purpose of the restoration of the beauty of the countryside which has been injured by certain works. It would appear that this law, which has in fact been out of use for some time, could be applied by associatons of a legal character and by physical individuals.

Although there are only limited possibilities for associations to act in their own name for compensation for certain ecological damage, in practice their role can nevertheless be of importance. It goes without saying that associations can stimulate and assist their members to claim a personal damage and to take legal action under their own name. Equally, they can represent their members on the basis of a proxy.[32]

12.1.1.2.3 Damage suffered by a public body

The position regarding public bodies is similar to that of associations.

The State, the provinces and the municipalities can obviously introduce actions for compensation when pollution has damaged their moral or material assets.[33] It is, however, not sufficient for the public body to claim that certain activities are contrary to their objectives. The violation of a law or a regulation, for example, does not in itself constitute a moral injury to the public body which issued the provision that has been violated. Part of the jurisprudence, hard to justify on principle, does, however, admit the constitution of a civil party by the muncipality following certain infringements of the legislation on land development and town planning, and grants it symbolic general damage interests.[34] In addition, it is clear that a municipality cannot claim individual injury suffered by its inhabitants to be personal damage.[35]

Certain special legislation enables the communes and certain other public bodies to act by civil means without the need to establish personal damage. This is the case in matters of environmental protection with regard to land development and town planning[36] and non-navigable water courses.[37]

12.1.1.3 THE CAUSALITY LINK

For a causality link to exist between an injury and a fault, it is sufficient that the injury, as it is presented in the particular case, would not have occurred without the fault.[38] It is therefore an essential condition of damage and is sufficient in itself. According to the theory widely accepted in Belgium, but not without nuances and hesitation on the part of the judges, of the equivalance of conditions, no differentiation is made between the various circumstances which led to the existence of the damage.[39]

In principle and by comparison with the theory of adequate causality, this approach leaves great possibilities for establishing a causality link between ecological damage and a source of pollution. No theoretical distinction is made between direct and indirect damage. This does not prevent the causality link from frequently being one of the essential obstacles in the outcome of an action for compensation for damage to the environment.[40] First of all, in certain specific cases, it would even appear difficult to satisfy the requirements of the theory of equivalence of conditions. Great importance lies in the difficulties regarding proof. Third, there is the situation where the same damage is the result of different sources of pollution. Finally, there are exonerating causes and, more particularly, the question of knowing whether the acceptance of risks, the prior presence of the polluter or the special susceptibility of the victim can be considered.

12.1.1.3.1 Certain special situations

In the case where ecological damage occurs in an environment which is already very polluted, it will be difficult to demonstrate that a specific source of pollution is a direct condition of that occurrence. This consideration will limit the effectiveness of a responsibility action, for example, in combating the almost generalised pollution of certain rivers.[41]

Similarly, there is a problem with regard to the recovery of costs on behalf on the exposed community in the prevention, research and neutralisation of the pollution[42] (costs of cleaning a beach or a road after an oil spillage). The application of general principles here results in a distinction between 'general' costs which it is certain that the State would not have incurred without the fault of the defendant, and those which would, even without this specific fault, have been incurred in the fulfilling of certain legal obligations (such as maintaining the roads). The Court of Cassation, however, decided on 28 April 1978 (with the application of an old idea) that the costs incurred by the complainant in carrying out a legal obligation cannot be recovered as compensation for damage.[43] The consequence, which is open to criticism, of this jurisprudence would be that the State or a muncipality would not be compensated by the polluter for the costs of the necessary safety measures resulting from a pollution incident which was the fault of the polluter.[44]

The practical importance of these rules of common law is, however, largely reduced by two legal provisions inspired by the 'polluter pays' principle. First, there is article 85 of the Law of 24 December 1976 'relating to the 1976–77 budgetary proposals'. The first paragraph of this article states that the State and the communes 'are required to recover' the costs caused to the civil protection services and the communal fire services for any intervention as a result of their legal and regulatory obligations in the event of 'duly verified contamination or pollution'.[44a]

These costs should be recovered from the 'owners of incriminating products' who, in accordance with the rules of common law, can take action against a responsible third party. Although this provision is not very explicit on this subject, it appears to impose an objective responsibility on the owner of products which have given rise to pollution. Equally important, but with a narrower field of application, are articles 16 and 18 of the law on toxic wastes which make the possessor of toxic wastes or the person who has committed the infringement responsible for the costs relating to the safety measures taken by the public services.[44b]

12.1.1.3.2 Proof

The essential difficulty for the complainant often lies in proving the causality link. It can, in fact, be very difficult—and costly—to localise the source of a certain pollution incident, as well as to demonstrate that this source is the origin of a specific damage.

In Belgian law the burden of proof (which can be supplied by any legal means, including supposition) rests on the complainant.

In this context, the discretionary powers of the judge in evaluating this proof are of great importance. It is said that the judges do not always show an equal severity in this respect.

In addition to decisions requiring very formal, scientific proof of the existence of a certain damage and its origin, there are others where the judge is easily satisfied with greater or lesser probability and leaves it to the polluter to supply proof to the contrary.[45]

Articles 871 and 877 of the Judiciary Code can be very useful to the complainant since they can require that the defendant or a third party exhibits certain elements of proof which are at their disposal.

12.1.1.3.3 Plurality of causes

In the event of several causes, the situation of the victim is noticeably improved, at least in theory. If the damage is the result of a fault which is common to different polluters, they are jointly responsible.[46] If, on the other hand, it is the result of a variety of different faults committed by several people, they are responsible *in solidum*. In both cases, all the polluters can be sued for the total damage.

However, there are relatively few joint and *in solidum* judgements in the jurisprudence regarding ecological damage. First of all, it is very rare that a pollution incident is the result of the *same* fault committed by several people. In the event of *in solidum* responsibility, on the other hand, it is necessary that the faults of the different polluters are all the cause of the *same* damage,[48] without it therefore being possible to identify the particular part of the damage caused by each source of pollution. This condition is achieved each time that different pollution incidents become harmful, for example, because they exceed the assimilating capacity of a river or have a synergistic effect. In the majority of published decisions on the subject, the judge succeeds in identifying the particular part of the damage belonging to each of the concurrent sources of pollution.[49]

12.1.1.3.4 Exonerating causes

Traditionally the most important exonerating causes, the unknown cause and the fault of the victim, do not raise particular problems on the subject of ecological damage.

The acceptance of risks by the victim, the particular susceptibility of a victim to damage and the prior presence of the polluter are other matters. These elements have frequently, in the past, been taken to partially or totally exonerate the polluter. In the context of responsibility for fault (and contrary to the objective theory of the destruction of the equilibrium) this result cannot as a general rule be justified unless the victim himself has committed a fault. A recent part of the jurisprudence has sometimes deviated from this rule.[50]

12.1.1.4 METHODS OF COMPENSATION

The principle objective of a responsibility action is to establish as far as possible the state the victim would have been in if the fault had not been committed.[51] This will preferably be done by compensation in kind through the removal of the consequences still in existence of the illegal act (restitution of objects removed, destruction of illegal works, injunction to carry out certain works).

If this kind of compensation in kind is impossible, financial compensation is paid. This will consist either of a lump sum or a form of rent, dependent on the case in question.

The power of the judge to forbid certain activities or to impose certain precautions to prevent future damage resulting from a repetition of the illegal harmful act is of great importance for the protection of the environment. These measures frequently resemble compensation in kind and are often mistaken for it. They nevertheless have distinct characteristics and can only be imposed if it is established with certainty that the faulty actions (even future) of the defendant would be harmful to the complainant.[52,53]

The effectiveness of the responsibility action in obtaining direct sanctions (prohibition, injunction) is limited by three elements.

First, the violation of a prohibition or an injunction by a judge is not sanctionable under the penal code. However, since the Benelux Convention on constraint[54] came into force, the judge is able to pronounce a judgement of constraint without having to justify it as compensation for future damage which would be suffered by the victim in the event of the decision not being respected.[55]

Second, the theory of the abuse of right enables the judge to limit himself to financial compensation in the case where the requirement of direct sanctions would constitute an abuse of right. The jurisprudence relating to ecological damage very rarely applies this possibility.[56]

A third restriction of the judge's power to pronounce injunctions and prohibitions is found in jurisprudence and traditional practice in the fact that the defendant exercises his polluting activities in conformity with an administrative authorisation. The dominant view on this subject, formulated by the decrees of the Court of Cassation of 27 April 1962 and 26 November 1974[57] can be summarised as follows:

(a) The courts are competent to recognise the damage caused by authorised establishments.

(b) The judge can impose precautionary measures to avoid future damage, but only provided that these measures are not in contradiction with those imposed by the administration in the general interest and that they do not endanger, even temporarily, the existence of the establishment referred to. This restrictive solution is generally justified on the basis of a division of power and of the fact that the authorisation is more than a simple police measure and also regulates the civil rights of the operator and his neighbours.

A considerable part of legal practice criticises these concepts and defends the position of the judge as having the same powers towards an authorised establishment as towards another establishment. A part of the jurisprudence has always refused to follow the restrictive concept.[58]

With regard to other authorisation procedures (building, works on non-navigable water courses, etc.) the problem does not arise: on these subjects the jurisprudence has generally recognised the power of the judge to impose direct sanctions.[59]

12.1.1.5 RESPONSIBILITY OF THE PUBLIC POWERS

The public powers have a considerable influence on the state of the environment. They themselves are in fact the cause of many environmental disturbances which they either approve or tolerate.

The general rule concerning responsibility for faults committed by the public powers[60] applies equally to compensation for the damage they cause.

In recent decades the jurisprudence has vigorously upheld the view that the public powers are subject to the law in the same way as the ordinary citizen. However, the discretionary powers of the public authorities greatly restrict the power of the judge in his evaluation as to whether or not their activities are in accordance with specific provisions of a legal or regulatory nature: the judge's legal control therefore ends at the point where these discretionary powers begin.

The discretionary powers of the public authorities also influence the judge's application of the general obligation for care and attention under article 1382. The jurisprudence of the Court of Cassation states that on this level the judge can fully exercise his legal control but he must not involve himself in questions of opportuneness. In legal practice[61] this dilemma has sometimes been resolved by suggesting that the judge restricts himself to a marginal control in those cases where the application of the obligation for care and attention implies a judgement on the manner in which the public power has assessed the different values in question, as a result of its discretionary powers; the judge cannot then declare the activity of the public power to be contrary to article 1382, except where this activity has been manifestly unreasonable. These restrictions scarcely apply when the application of article 1382 does not imply this value judgement on the manner in which the government has exercised its discretionary powers. An integral application of the standards implied by the obligation to respect the state of technology and the state of the art is therefore still possible.

As a general rule, the role played by this discretionary power in decisions taken by public authorities which have ecological impact is very great. This is more particularly the case where the following factors are involved: public works, the production and distribution of energy, means of transport, defence, treatment plants, the exercising of regulatory power, permit approval and the exercising of police control.[62] On all these subjects, adhering to the concept described above, the judge must limit himself to a marginal control so that the activities of the public powers are in accordance with the rules for care and attention. Moreover, an examination of the jurisprudence reveals only very few examples where the actions of the public powers which have had damaging consequences for the environment have been judged to be at fault because of their intrinsic value.

Of greater interest are the rules on competence, form and procedure surrounding the public authorities, with the growing number of technical regulations, as well as rules on care and attention such as the obligation to respect the state of technology or the state of the art. The application of these is not influenced by any discretionary powers. Their violation of a public power constitutes a fault in the same way as when a private individual is involved.

In the event of faulty procedure or failure by the public powers to act in accordance with the regulation or to exercise sufficient police control being invoked as faults, an action for compensation would be limited because of the difficulty of establishing a causality link between these faults and ecological damage.

On the sanction level, the responsibility of the public powers also presents several special and important characteristics for compensation for damage to the environment.

On the basis of the division of power, the tribunals traditionally hesitate before pronouncing an injunction or prohibition against a public body.[63] However, in recent jurisprudence and practice a fairly precise move has been established in favour of the application of direct sanctions to government.[64] This evolution is of considerable importance in the fight against disturbance to the environment and it is fully justified on the theoretical level.

It should be noted that the ability of the judge to impose direct sanctions is nevertheless limited by the discretionary powers of the public authority. As a result, the judge cannot prescribe the means of achieving the objectives aimed at by the legislators.

Once again, there is an absence of secondary sanctions which would force the public powers to observe the injunction or prohibition. The approval of the uniform law on constraint (which provides for the possibility of imposing constraint on the public powers) partially remedies this problem. Nevertheless, in the opinion of most authors on the prevailing jurisprudence, implementation by force is not possible against a public body.[65]

12.1.2 Liability arising from defects in objects

12.1.2.1

Article 1384 (Civil Code), according to Belgian interpretation, contains an irreversible assumption of fault on the part of a person in charge of an object, the defect of which has caused damage to others. This provision is of a certain limited importance in compensation for ecological damage.

The application of article 1384(1) is in fact surrounded by fairly strict conditions.

First, it is only applicable to material objects. Faults in the management or planning of an industrial complex, for example, do not lead to the application of article 1384(1).[66] For this, a defect in one of the material elements of the complex must be established.

The principle factor which determines the sphere of application of article 1384(1) is the interpretation given to the idea of defect. According to recent jurisprudence, this must involve an intrinsic and abnormal characteristic which would make the object dangerous.

It is therefore not sufficient for an object to be dangerous because of its nature. The danger must result from an abnormality.[67] The fact alone that an industrial installation creates a nuisance is not therefore sufficient for it to be considered defective: this is the case in the majority of installations.[68]

The defect must to some extent be inherent in the object. The unexpected and dangerous modification (explosion, combustion) of an object is not considered to be defect[69] by the Court of Cassation which has also refused to consider the fact that an object is found in an abnormal place to be an external defect.[70]

It is important to note that the severity of these concepts is partially mitigated by the possibility of providing negative proof of the defect: it is sufficient to establish that no cause other than a defect can reasonably explain the damage.[71]

All the implications of the jurisprudence on the subject of the theory of defect are still unclear. It can, however, be concluded that a rubbish dump is not defective because it catches fire and a purification plant is not defective merely because it does not produce the expected result.[72]

On the other hand, article 1384(1) is applicable to environmental damage caused by an anomaly in material components or in their mutual relationship to an object. Its interest lies above all at the level of accidental pollution due to technical failures in industrial installations (the breakage of a pipe,[73] the breakdown of a machine[74]).

12.1.2.2

The sanctions involved in the application of article 1384(2) are broadly the same as those relating to article 1382. Financial compensation is therefore a possibility, as is compensation in kind.

Since the use of an object which is known to have caused damage as a result of being defective constitutes a fault under article 1382, if necessary, it is also possible to obtain an injunction against the person in charge of the object to put right the defect.[75]

12.1.2.3

The application of article 1384(1) to the public powers does not give rise to specific problems. Without hesitation the jurisprudence has declared the organisations of the public powers to be responsible for damage caused by defects in objects in their possession.[76]

12.2 STRICT LIABILITY

In addition to subjective responsibility based on article 1382 and subsequent articles, the jurisprudence has, in adopting the theory of the destruction of a balance between neighbouring properties, introduced a regime of strict liability, the importance of which is particularly significant with regard to compensation for damage to the environment.

In recent years, the legislator has also introduced strict liability for damage resulting from certain well determined injuries to the environment.

Mention should also be made of article 7 *bis* of the Law on the Council of State which enables it to grant compensation for certain exceptional forms of damage resulting from the activities of an administrative authority and for which no compensation can be obtained on the basis of ordinary law.

Finally, the extent to which there exists a *collective* responsibility for environmental damage should be examined.

12.2.1 Liability for upsetting the balance between neighbouring properties

Confronted with the theoretical difficulties involved in the application of the idea of fault to certain more or less inevitable nuisances, in 1960[77] the Court of Cassation adopted the theory of the disturbance to the balance between neighbouring properties elaborated by the doctrine. The Court decided that anyone committing an act which, though not a fault, disturbs the balance (established between neighbouring properties), by exposing the owner of a neighbouring property to disturbance exceeding the normal inconveniences of neighbourliness, owes him a just and adequate compensation to restore the disrupted balance. This new responsibility, the basic theory of which has been widely criticised,[78] is firmly established in the jurisprudence. It is of particular importance for damage to the environment and differs on several points from the responsibility for fault. There must be a neighbourly relationship between the complainant and the defendant for it to be invoked, and the damage must exceed the normal inconveniences of neighbourliness. Finally, the sanction is not merely compensation for the damage, but a just and adequate compensation to restore the disrupted balance.

12.2.1.1 THE RELATIONSHIP BETWEEN THE COMPLAINANT AND THE DEFENDANT

Although the judgements of the Court of Cassation confined the application of the new theory to the relationship between neighbouring properties, this restrictive concept was soon abandoned. With the approval of the doctrine, the jurisprudence no longer actually requires *owners* to be involved, but will accept neighbours with the right to use or enjoy the property.[79] The theory of the disturbance of the balance is therefore applicable as soon as someone such as the owner, lessee, user or holder of another real right to the enjoyment of a property suffers damage in excess of normal neighbourly inconvenience as a result of activities carried out on the neighbouring property.

The idea of neighbourliness is widely interpreted. No contiguity is required. According to some decisions it is enough for a property to suffer from the effects of activities carried out on another property.[80]

The theory of the disruption of the balance is therefore widely applicable to ecological damage caused by fixed sources of pollution such as industrial establishments and construction yards.

12.2.1.2 DISTURBANCES EXCEEDING THE ORDINARY INCONVENIENCES OF NEIGHBOURLINESS

12.2.1.2.1

Responsibility for disturbing the balance can only be invoked if the damage exceeds normal neighbourly levels. Damage exceeding the normal is essentially relative. The practical content is largely determined by the degree of industrialisation and urbanisation of the region where the disturbance occurs, as well as by the state of technology.

Moreover, the judge has great power in evaluating the situation.

The theory of neighbourly disturbance has often been applied to material damage resulting from public or other construction works. It has also been successfully invoked against different forms of industrial and other nuisances.[81]

The idea of abnormal disturbance or of disturbance exceeding the normal neighbourly limits of tolerance, while offering considerable possibilities of application to the subject of environmental damage, is nevertheless questionable from the point of view of juridicial policy. It implies that whatever is usual is acceptable and thus contributes to the perpetuation of certain types of ecological damage.[82]

12.2.1.2.2

Contrary to the situation regarding the responsibility for fault, the abnormal susceptibility of the victim and the prior occupation of the polluter are called upon to play a fairly considerable part in the context of responsibility for disruption of the balance.[83] Abnormal disturbance is, in theory at least, an objective idea, the content of which depends on a comparison between the nuisance resulting from the activities of the defendant and the normal levels of neighbourly inconvenience.

This objective character explains why there can be responsibility for damage which the victim feels to be excessive, only because of his abnormal susceptibility.

It is obvious that collective concern (the general character of the region) will also be taken into consideration. As already mentioned, the idea of abnormal disturbance precisely implies a reference to the character of the region.

On the other hand, individual concern (the fact that the polluter was established in a certain place before the complainant) does not normally have any influence on the responsibility for disrupting the balance, any more than it does in the context of article 1382. The fact of being established in a certain place before the neighbours were there, does not mean that the normal tolerance level of the neighbourhood is greater. This would only be the case if the polluting establishment contributed to the determination of the character of the region. However, part of the doctrine does grant a greater role to individual concern.[84]

12.2.1.3 COMPENSATION

12.2.1.3.1

On the level of sanctions, the theory of disruption of the balance differs from fault liability in that an injunction or prohibition aimed at controlling the source of pollution can be obtained. Works can only be implemented on the property of the defendant in order to protect him. On the other hand, financial compensation for the damage incurred largely covers the same losses as article 1382 c.c.[85]

The impossibility of asking for direct sanctions appears to be inherent to the theoretical basis of the idea of the disruption of the balance. It is not possible to prohibit activities which are neither illegal nor faulty. The arguments of juridical policy invoked to justify the impossibility of an injunction or prohibition are, however, not convincing. A solution which would allow more varied sanctions would be preferable.

12.2.1.3.2

The application of the theory of disruption of the balance to the public powers scarcely raises practical problems, even though the jurisprudence and the doctrine have a tendency to invoke the theory of the equality of citizens and public powers, of which the practical results are the same.

12.2.2 Special regimes of strict liability

12.2.2.1

The Belgian legislator has introduced a special regime of responsibility without fault for various forms of pollution.

This applies to toxic wastes and the exploitation of groundwaters, as well as to nuclear energy and marine pollution by hydrocarbons.[87]

12.2.2.2

The Law of 22 July 1974 imposes on the producer of toxic wastes an objective responsibility for all damage caused by these wastes, whether the damage occurs during the course of their transport, destruction, neutralisation or disposal (article 7).

In the case where the governor of the province makes use of his power to impose conditions, seize, destroy, neutralise or dispose of toxic wastes which have been dumped or handled in violation of the law, the person who committed the infringement is responsible for the costs incurred for this purpose (article 16). If the governor or the burgomaster orders the transport of certain toxic wastes because their presence causes serious danger, the person who was in possession of the wastes bears the cost of this operation (article 18).[88]

The law also provides for the creation of a 'Guarantee Fund for the Destruction of Toxic Wastes' (article 9), financed by contributions from businesses producing, handling or treating wastes (article 12). In the event of bankruptcy or insolvency of those responsible, the fund ensures that their obligations as described above are met (article 11). The administrative council of the fund can decide to intervene in other situations where the person responsible remains at fault (article 11).

12.2.2.3

The Law of 10 January 1977 organises compensation for damage caused by the off-take and pumping of groundwaters. Under the terms of the first paragraph of this law, the operator of a groundwater off-take and the head of the public or private works which, as a result of activity, cause a drop in the groundwater level, are objectively responsible for the resulting surface damage to immovable property, including soil and vegetation, or for damage caused to existing groundwater offtakes of more than 10 m^3 per day. The law does not apply to damage resulting from pump works in mines or quarries.

Since responsibility for disruption of the balance can already be applied to situations covered by the law, it does not bring about any fundamental changes to the rules already in force.[89] Furthermore, it does not remedy the major difficulty for the victims which consists in establishing the causality link between the damage, a lowering of the water level and certain water pumping operations.[90]

A more significant development than the Law of 10 January 1977 is the creation of an 'Advance Fund for the Compensation for Damage Caused by Off-takes and Pumping of Groundwaters' (article 7). The purpose of this fund is to provide the victim with a financial advance when a summary inquiry has established the existence of a connection between the damage, a lowering of the water table and an offtake or pumping of water (article 8). It is important to underline that this is only an *advance*. If the responsibility action is finally dismissed by the judge, the fund recovers the advance paid to the complainant without his being able to claim interest (article 8, s.2). The essential benefit provided by the fund is therefore that it enables compensation for damage to be speeded up.

12.2.3 Exceptional damage

Article 7 *bis* of the Law of 23 December 1946 states that the administration section of the Council of State should decide on the fairness, taking into account all the circumstances of public and private interests, of requests for compensation relating to an exceptional moral or material damage caused by an administrative authority.

This little-known provision of the ordinary responsibility laws is rarely applied and is only mentioned here in order that a complete picture be given since it can apply to certain types of ecological damage.

There are numerous restrictive conditions placed on the Council of State's approval of compensation requests for exceptional damage. It must be exceptional and by its nature or size must exceed the irritations and sacrifices normally incurred in everyday life and must therefore constitute a violation of the citizen's right to equality in the face of the public powers.[92]

No other jurisdiction may be competent to evaluate the request for compensation; all damage caused as a result of a fault, or for which an objective responsibility exists under civil law, is therefore excluded. The demand is only finally acceptable once the administrative authority has totally or partially rejected a request for compensation, or has failed to reach a decision on the matter within sixty days.

Even when all those conditions have been satisfied, it is not certain that the Council of State will grant the compensation requested: it will decide on the fairness of the request, taking into account all the circumstances of public and private interest and, in particular, the financial burden which the compensation would present for the public authority.

12.2.4 Collective responsibility

As a general rule, objective responsibilities for ecological damage in Belgian law are individual responsibilities.

There have, however, been several cases where compensation for certain damage to the environment has been made collective.

Reference can be made here to the intervention in the event of bankruptcy or insolvency by the Guarantee Fund for the Destruction of Toxic Wastes[93] which is collectively financed by the businesses which produce or treat toxic wastes.

Apart from this, there is only the social legislation on the basis of which it would be possible to obtain financial compensation for certain types of damage resulting from injury to health caused by the degradation of the environment or by any other cause.

In the context of compulsory health and disability insurance,[94] financial compensation can be obtained to cover the costs of medical expenses and loss of earnings. Where it is possible to identify the polluter and to establish his responsibility, according to the rules of ordinary law, the insurance organisation, substituting for the victim, can recover any compensation paid. Where this is not the case, compensation for damages incurred as a result of the pollution will be financed by the bulk of social contributions paid by employees and employers.

Where the damage constitutes an accident at work,[95] the victim would be entitled to certain compensation payments from the approved insurance organisations with which employers are obliged to insure their personnel against accidents at work. The Fund for Accidents at Work intervenes in the event of non-insurance or the insolvency of the insurer. The insurer who has paid the victim substitutes for the rights that the victim can exercise against any third party responsible for the accident or against an employer, his employees or representatives who intentionally caused the accident. In cases where this does not apply, accidents at work resulting from pollution incidents are then underwritten by the insurers and financed by the premiums paid by employers.

A victim suffering from an illness which is recognised as being an occupational illness[96] is entitled to compensation from the Fund for Occupational Illness, which is partly financed by contributions from employers and partly by State subsidy. This fund substitutes for the rights that the victim can exercise against a responsible third party, or against any employer, his employees or representatives who have intentionally caused the accident. When this is not the case, compensation for an occupational illness resulting from a pollution phenomenon is paid collectively.

Notes

1. Cass., 28 June 1974, *Pas.*, 1974, I, 1113. Cass., 30 April 1976, *R.W.*, 1976–77, 1709.
2. Cass., 31 January 1944, *Pas.*, I, 1944, 178. Cass., 13 February 1970, *Pas.*, 1970, I, 511. R. O. Dalcq, *Traité de la responsabilité civile,* I (2nd ed), Brussels, Larcier, 1967, no. 301, p. 178.
3. E.g. Cass., 6 December 1965, *Pas.*, I, 450.
4. Cass., 26 September 1955, *Pas.*, 1956, I, 34. Dalcq, *op.cit.*, I, no. 304, p. 179.
5. Law of 30 May 1961.
6. H. Bocken, *Het aansprakelijkheidsrecht als sanctie tegen de verstoring van het leefmilieu,* Brussels, Bruylant, 1978, pp. 42–43.
7. H. Bocken, *op.cit.*, pp. 43–48.
8. Comm. Courtrai, 31 March 1978, R.G. no. 249/77, n.p.; Ref. Anvers 5 May 1978, n.p.
9. Liège, 8 October 1969, *R.G.A.R.*, 1968, 8341; Liège, 13 January 1976, R.G. no. 5.681/75, n.p.
10. Comm. Courtrai, 31 March 1978, R.G. no. 249/77, n.p.
11. Corr. Gand. 23 July 1958, *B.J.* 1858, 982; Gand, 9 February 1877, *Pas.*, II, 132; Ref. Anvers, 5 May 1978 and 18 May 1878, n.p.
12. H. Bocken, *op.cit.*, pp. 66–67.
13. The application of this interpretation of the idea of the abuse of right in matters of pollution is defended by A. Berenbook, 'Chronique de jurisprudence, Droits réel', *J.T.*, 1974, 275, no. 12.
14. *Arr. Verbr.*, 1972, 32.
15. Corr. Verviers, 24 December 1968, *J.C.P.* (Ujr), 1970, 16535, Note Despax; Gand, 9 December 1976, n.p.
16. Civ. Anvers, 26 March 1937, *R.W.*, 1937–38, 205; Civ. Anvers, 23 April 1959, *R.W.*, 1959–60, 787.
17. Cass., 16 January 1939, *Pas.*, 1939, I, 15; R.O. Dalcq, *op.cit.*, II (1st ed, 1962), no 2915, 286.
18. E.g. Brussels, 19 November 1965, *R.W.*, 1965–66, 950; Gand, 25 November 1969, *J.T.*, 1970, 205.
19. R. Derine and P. Hamelinck, 'Overzicht van rechtspraak (1969–1972) Zakenrecht', *T.P.R.*, 1973, 798; H. Bocken, *op.cit.*, pp. 106–111.
20. E.g. Pol. Brux. 21 October 1957, n.p.: '. . . that the civil party has been manifestly inconvenienced by the enormous cloud of black smoke arising from taxi exhaust fumes . . . that the guilty act has incontestably led to a slight blow to the physical wellbeing of D. . . ., in respect of which he has a right'.
21. See in this context the judgements regarding obstruction to the use of a by-road: Cass., 19 December 1895, *Pas.*, 1896, I, 48; H. Bocken, *op.cit.*, pp. 89–91.
22. On these points see H. Bocken, *op.cit.*, pp. 84–92.
23. The situation can be different for actions cancelling acts of the public powers brought before the Council of State (Mast., *Administratif Recht,* no. 546, 515).
24. H. Bocken, *op.cit.*, pp. 100–101; Cass., 6 December 1977, *R.W.*, 1977–78, 1688. This possibility does not exist in the procedures before the Council of State.
25. E.g. Corr. Gand, 16 March 1973, *Entr. et Dr.*, 1973, 233.
26. H. Bocken, 'De bevoegdheid van milieuverenigingen tot het instellen van burgerrechtelijke vorderingen tot sanctionering van de verstoring van het leefmilieu', *Entr. et Dr.*, 1973, 239.

27. E.g. Civ. Ypres, 6 June 1963 and Gand, 21 December 1963, *Entr. et Dr.*, 1973, 230; Corr. Verviers, 24 December 1968, *J.C.P.* (jur) 1970, 16535; Corr. Hasselt, 18 April 1972, *R.W.*, 1972–73, 775.
28. E.g. Cass., 28 May 1934, *R.G.A.R.*, 1935, 2022.
29. Cass., 9 December 1957, *R.C.J.B.*, 1958, 247; Cass., 28 June 1968, *Pas.*, 1968, I, 1239.
30. E.g. Gand, 9 December 1976, R.G., 35188, n.p.; Corr. Huy, 15 June 1978, no. 332, G.R. E.g. Ref. Anvers, 29 September 1976, R.W., 1976–77, 560. A liberalisation has also intervened in the jurisprudence of the Council of State (11 February 1977, no. 18.101. A.s.b.l. Association for the Preservation of the Environment of Nivelles, 30 June 1978, no. 19.114, a.s.b.l. Interenvironment wallonie).
31. Cass., 9 July 1928, *Pas.*, 1928, I, 227.
32. Cf. note 14 (2).
33. E.g. the action introduced by the commune of Houthulst as a result of the pollution of one of its industrial sites (Gand, 25 November 1968, *J.T.*, 1970, 205).
34. H. Bocken, *op.cit.*, p. 104, note 279.
35. Corr. Gand, 23 July 1858, *B.J.*, 1858, 982.
36. Article 67(1) of the Law of 29 March 1962.
37. Royal Decree of 5 August 1970, article 15.
38. Cass., 19 June 1967, *Pas.*, 1967, I, 1274.
39. See R. O. Dalcq. *op.cit.*, II, no. 2352 and following, no. 102 and following and no. 2392 and following, no. 114 and following; Van Quickenborne, *De oorzakelijkheid in het recht van de burgerlijke aansprakelijkheid,* Gent, 1972, no. 404, 231.
40. See H. Bocken, *op.cit.*, pp. 112–140.
41. E.g. Civ. Brussels, 24 April 1869, *B.J.*, 1869, 936.
42. H. Bocken, *op.cit.*, p. 116.
43. *R.W.*, 1978–79, 1695. (It was a question of the expenses incurred by the town of Antwerp to remove a wreck from the Escaut. The town was regulatorily obliged to carry out these works if the owner failed to do so.)
44. Cf. the law on toxic wastes which introduces an objective responsibility for damage of this type (see no. 40).
44a. The Royal Decree of 23 January 1971 imposing on the civil protection services the obligation to intervene in the event of 'catastrophes: particularly . . . contamination and pollution by an accident of a certain seriousness or affecting a large regional area in a general manner, and all accidents which involve or could lead to the loss of numerous human lives or considerable material damage'; the first paragraph of article 85 of the Law of 24 December 1976 would normally apply in cases of serious pollution.

As a result of the second paragraph of article 85 of the Law of 24 December 1976, the costs incurred by the civil protection services and the communal fire services during operations carried out by these services, *outside the intervention they are obliged to carry out under the laws and regulations,* must be recovered by the State and the communes *from the beneficiary of such operations.*

This provision can apply when the intervention of the safety services is requested following an incident of pollution which is not sufficiently serious to merit an emergency intervention on their part.
The Royal Decrees of 27 January 1978 and 9 August 1979 control the methods of evaluation and recovery of the costs.
44b. See note 39.
45. Comp. Civ. Nivelles, 13 May 1970, *Rec. Niv.*, 1971, 98 and Civ. Gand, 29 June 1977, n.p.
46. Cass., 15 February 1974, R.W., 1973–74, 1715 and conclusions of the Solicitor-General, F. Dumon.

47. Cass., 9 October 1975, *Pas.,* 1976, I, 43.
48. Cass., 4 April 1947, Pas., I, 956.
49. H. Bocken, *op.cit.,* pp. 126–128.
50. H. Bocken, *op.cit.,* pp. 129–140.
51. Cass., 2 May 1974, *Pas.,* 1974, I, 906; R. O. Dalcq, *op.cit.,* II, no. 4139, 741.
52. H. Bocken, *op.cit.,,* pp. 144–151.
53. These preventive measures cannot be imposed by the penal judge, who can only take account of the civil consequences of an infringement already committed (Bocken, *op.cit.,,* p. 151, note 483).
54. Treaty of 26 November 1973.
55. Law of 31 January 1980, *Moniteur belge,* 23 February 1980.
56. Civ. Liège, 11 January 1960, *J.L.,* 1950–60, 206; H. Bocken, *op.cit.,* pp. 156–157.
57. *Pas.,* 1962, I, 938 and *R.W.,* 1974–75, 1768.
58. H. Bocken, *op.cit.,* pp. 157–169.
59. H. Bocken, *op.cit.,* p. 161.
60. See J. Delva, 'Civielrechtelijke aspecten van de overheidsaansprakelijkheid', *R.W.,* 1977–78, 2343.
61. H. Bocken, *op.cit.,* p. 189.
62. H. Bocken, *op.cit.,* pp. 191–239.
63. M. A. Flamme, 'Pour un contrôle juridictionnel plus efficace de l'administration', *J.T.,* 1972, 417.
64. H. Bocken *op.cit.,* pp. 342–346.
65. K. Baert, 'De uitvoeringsimmuniteit van de publiekrechtelijke rechtspersonen', *R.W.,* 1976–77, 2369.
66. Cass., 21 April 1971, *R.W.,* 1972–73, 567 (hospital establishment).
67. Cass., 23 September 1971, *Pas.,* 1972, I, 80.
68. H. Bocken, *op.cit.,* p. 258.
69. Cass., 24 December 1970, *R.W.,* 1970–71, 1470.
70. Cass., 27 November 1969, *R.C.J.B.,* 1970, 41.
71. Cass., 25 March 1943, *Pas.,* 1943, I, 110.
72. H. Bocken, *op.cit.,* p. 260.
73. Mons, 18 November 1975, *Pas.,* 1976, II, 136.
74. Liege, 13 January 1976, R.G. no. 5.681/75, n.p.
75. H. Bocken, *op.cit.,* p. 263.
76. Cass., 9 November 1972, *R.W.,* 1972–73, 1953.
77. Cass., 6 April 1960, *Pas.,* 1960, I, 920.
78. H. Bocken, *op.cit.,* pp. 270–275.
79. Cass., 10 January 1974, *Pas.,* 1974, I, 520.
80. Civ. Liège, 25 February 1969, *Entr. et Dr.,* 1971, 225.
81. E.g. Cass., 19 October 1972, *Arr.Verbr.,* 1973, 178.
82. H. Bocken, *op.cit.,* p. 287.
83. H. Bocken, *op.cit.,* pp. 287–290.
84. M. A. and Ph. Flamme, 'Chronique de jurisprudence. Les troubles de voisinage', *Entr. et Dr.,* 1974, 197.
85. H. Bocken, *op.cit.,* pp. 291–296.
86. Brussels, 11 February 1970, *Entr. et Dr.,* 1974, 107; Gand, 20 April 1972, *Entr. et Dr.,* 1972, 217; M. A. Flamme, *Traité des Marchés Publics,* Brussels, 1969, II, 751.
87. An occasional interest for compensation for environmental damage is to be found in the legislation on mining damage (Royal Decree of 15 September 1919, article 50), the Law of 10 March 1925 on the distribution of electricity (articles 14–17) and the Law of 12 May 1927 on military requisitions which was invoked to hold the Belgian State objectively responsible for damage caused by the noise of military aeroplanes (civ. Nivelles, 21 January 1969, *Rec.Jur.Niv.,* 1970, 189; M. Litvine, 'La responsabilité pour dommages dérivant du bruit des détonations balistiques provoquées

par les aéronefs', in *Rapports belges au VIIIème Congrès de droit comparé*, Brussels, 1970, 494).

88. See note 20.
89. M. Hantoiau, 'La loi du 10 janvier 1977, organisant la réparation des dommages provoquées par des prises et des pompages d'eau souterraine', *Entr. et Dr.*, 1977, 193.
90. Idem, 195.
91. A. Mast, *Administratief Recht*, Gent, 1977, no. 595,563.
92. R. O. Dalcq, *op.cit.*, I, no. 1538, 506.
93. A comparable role is played by the Guarantee Fund for Coal-mining Damages (Royal Decree of 3 February 1961).
94. Law of 9 August 1963.
95. Law of 10 April 1971.
96. Laws relating to the compensation for damage resulting from occupational illness co-ordinated on 3 June 1970.

13

Conclusions

13.1

In Belgium, as elsewhere, developers and the public at large have not seen the advantages of environmental protection, or its relevance in everyday life, and they are therefore not aware of any urgent need to pursue such objectives.

This fact is reflected in the evolution of environmental legislation in Belgium.

While the control of dangerous, dirty and noxious establishments dates back to the 19th century (Imperial Decree of 15 October 1810), its objectives were very limited: to avoid or reduce damage and nuisance beyond normal neighbourly inconvenience—above all, from the point of view of the protection of private property.[1]

Preoccupations with habitat and town planning have evolved only during the last thirty years; this has resulted in a second stream of legislation, in particular the organic law on land development and town planning.

Finally, it is only very recently that the legislator in the environmental field has adopted a series of fragmentary and sectoral laws, each with its own distinct objective and its own structure.

13.2

This study of the juridical means of combating pollution in Belgium is certainly not complete. In addition to the legislation mentioned, there is a multitude of other legislative and regulatory texts aimed, at least indirectly, at preventing and controlling damage to the environment.

In particular, the Law of 7 August 1931 on the protection of monuments and sites and the regional decrees on the subject, the Law of 22 July 1970 relating to the restoration of the countryside, and the Law of 12 July 1973 regarding the protection of nature should be mentioned.

13.3

The leading role in these secondary texts[2] is undoubtedly played by the *organic law on land development and town planning*.[3]

The first article, clause 2, of this Law of 29 March 1962 states that land development should be conceived from an economic, social and aesthetic viewpoint as well as aiming to preserve the natural beauty of the countryside.

The different levels of plans envisaged are dependent on zoning, i.e. 'on the determination of zones for particular uses as defined by the public authority'.[4] The obligatory power of development plans implies not only that the authority which has adopted the plan in question must conform to it, but also that no authority can grant a permit for land use which is not authorised under the plan.

The regional executives and, within the limits of their competence, the communal councils, can issue general regulations on buildings and building plots. The regulations on buildings can contain provisions to ensure the cleanliness, solidity and appearance of the construction, the installation and its surroundings, as well as its safety.

Individual action is subject to an authorisation procedure for building permits and building plot permits. A prior written permit from the College of Burgomaster and Aldermen is required for building, demolition, reconstruction, deforestation, alteration of the landscape, etc., as well as the sale, or lease for a period exceeding nine years, of a plot forming part of a building plot for the erection of dwellings.

In addition to penal sanctions, sanctions intended to eliminate the effects of infringements can be imposed by the courts. These special sanctions can take three forms: the restoration of the site to its former state; the carrying out of development work; the payment of a sum representing the added-value acquired by the property following the infringement.

Numerous authors have pointed out the relationship between the environment and land development. Jacques Hoeffler, who has discussed these problems on several occasions, recalls justly that 'the legislation on land development and town planning is that which naturally has the most effect in the formation of a harmonious living environment'.[5]

13.4

M. Lamarque opens his important work on environmental legislation in France with the statement: 'Our legislative and regulatory arsenal is already well provided with provisions of all kinds to ensure the effective protection of nature and the environment; it is merely the desire to respect this legislation which is lacking'.[6]

The same applies in Belgium. It is undeniable that damage to the environment is reaching acutely worrying heights. What are the causes?

13.5 THE INSTITUTIONAL FRAMEWORK

The administrative organisation responsible for the implementation of laws controlling pollution must operate with the utmost effectiveness. Various factors come into play here. The technical nature of the problem, the need for scientific research and the lack of sufficient qualified staff are all factors which lead to centralisation. On the other hand, the desire of the public at large to participate in the decision-making process leads to decentralisation. As a result of these two opposing pressures, an institutional framework has to be set up.

In Belgium, the problem of the division of responsibilities is particularly complicated as a result of progressive regionalisation. During the last ten years the reform of political and administrative structures has resulted in all concrete action foundering in futile discussion over the respective powers of the national State and the regions.

At this time, the transfer of responsibility to the regions is based on the laws of 8 and 9 August 1980 regarding institutional reforms—see Chapter 1. 'The Belgian solution to this problem, which is naturally a compromise solution ... one can only regret the numerous 'exceptions' or 'reservations' of responsibility which have been introduced for the benefits of the national State regarding matters which have been transferred as a whole to the regions'.[7]

The distribution of responsibilities between the central administration and the regional authorities on the one hand and the division of responsibility between various ministries and ministerial departments and the national and regional plan on the other hand make it impossible to envisage the rapid development of a global policy on the environment.

But good environmental legislation presupposes a comprehensive vision of the qualitative objectives to be pursued.

13.6

Divergent national solutions would inevitably lead to distortions in competition. An effective programme of action on the environment must operate not only at national level, but also at European Community level.

In this context, the failure of the Belgian government to introduce directives on the environment into internal legislation should be recognised.

This failure is also explained by the change in the distribution of responsibility between central and regional government.

The Court of Justice of the European Communities judged that 'while they can explain the difficulty in implementing the directive, these circumstances do not hide the failure for which the Kingdom of Belgium is to blame. According to the constant jurisprudence of the Court, a Member State may not present practical arrangements or situations of an internal juridical nature as excuses to justify the failure to respect obligations resulting from community directives' (decrees of 2 February 1982).

It is doubtful whether the distribution of responsibilities as it exists at present could enable adequate measures to be taken to implement directives within the periods envisaged.

13.7

It appears from this study that Belgian legislation in the fight against pollution is scattered among legislative texts, with frequent recourse to the technique of outline laws, and there is no comprehensive law concerning the environment as a whole.

It is also true to say, as Baron Constant observed:

> In proceeding by an exclusively sectoral approach to control the different aspects of the environment, the legislator has constructed a work with a fatal lack of cohesion, with regard to the narrow interferences which the different forms of pollution present.

> On the other hand, certain laws have been stricken by sterility over months or even years, either because their implementing decrees have been issued late, or because the public authorities have not been

provided with the material means necessary to carry out the reforms envisaged.

Finally, in certain cases, the insufficiency of the punishments envisaged by the legislator—in particular the very low maximum sentences of imprisonment and the possibility of reducing punishments through extenuating circumstances—removes the intimidating character of the penal sanctions which play an essential role in the prevention of infringements.[8]

13.8

There are numerous monitoring measures available: automatic sampling, metering, notification obligation, etc. Is it necessary to underline that the existence of an effective monitoring apparatus constitutes an essential condition of the implementation of existing legislation?

Here again, enthusiasm is tempered by the means available. As an example we would quote the Food Inspection Service: in 1980 the 14 food inspectors and 27 controllers only drew up 152 infringement reports of which only a proportion gave rise to proceedings. It is useless to promulgate laws which exceed the capabilities of the State in terms of personnel and finance.

This is a general problem: the legislator votes good laws with great ease, leaving their implementation shrouded in uncertainty by failing to provide sufficient means for the departments concerned. This situation is doubly frustrating because it creates hopes which cannot be honoured.

13.9

What are the prospects for the future? Unfortunately, it must be admitted that they are not good.

There is just one glimmer of hope: in the Flemish Region, a comprehensive waste management policy seems to be well and truly at the forefront of political concern.

As for the rest, we must agree with Jean Untermaier: 'Legislative and regulatory inflation is doubling ... with an institutional proliferation: a problem, a text but also a technical committee, an interministerial committee and new actors regularly appearing on the administrative scene.'[9]

Notes

1. A. Ch. Kiss, The control of activities which are prejudicial to the environment by prior authorisation or declaration procedures, in: *Current tendencies in policy and law on the environment,* IUCN, 1980, p. 83.

2. See P. Dominice, The protection of the neighbourhood and the environment, *Ass. H. Capitant,* Paris, 1979, p. 365.

3. *Moniteur belge,* 12 April 1962.

4. Lilian Voye, Land development and land policy: sociological approach, *Rev. jur. ec. urbanisme et environnement,* Seres, April 1979, p. 13.

5. Jacques Hoeffler, The spatial dimension of the environment, in: *Juridical aspects of the environment,* Namur, 1975, p. 191.

6. J. Lamarque, Law on the protection of nature and the environment, Paris, 1973, p. XV.

7. Jean-Marie R. Van Bol, The environment, completely a regional matter? *Rev. jur. ec. environn. urban,* Seres, October 1980, pp. 8–9.

8. Baron Jean Constant, The penal protection of the environment in Belgian law, in: *Rapports belges au Xe Congres international de droit compare,* Budapest, 23–28 August 1978, p. 602.

9. Jean Untermaier, Law on the environment. Reflections for a first balance sheet, in: *Année de l'Environnement,* vol. I, *Revue du C.E.D.R.E.,* Paris, 1981, p. 24. It should be noted that Jean Untermaier refers to French law as a 'jungle of provisions' (p. 16).

Classified Index*

The Constitution, Public Authorities, Special Interest Groups and Individuals

The national constitution	1.1
Sources of laws governing pollution control and remedies for damage caused by pollution	1.1.3, 1.1.13
Government departments and agencies with supervisory, administrative or executive powers of pollution control	1.1.6, 1.2
National, regional and local public authorities with powers of pollution control	1.1, 1.2
Independent advisory bodies with rights or duties under pollution control legislation	1.3
Special interest groups representing those who may be liable for pollution, or those concerned to prevent or reduce pollution	1.4
Standing to sue *(locus standi)* in legal proceedings for pollution	2.3.4, 2.4.2

Air

Stationary Sources

Control by land use planning	2, 3.4.2, 3.4.4
Controls over plant and processes (including raw materials, e.g. fuels)	3.2.2, 3.3.1
Controls over treatment before discharge, and over manner of discharge (e.g. height of chimney)	3.3.2
Limits on emissions	3.3.2, 3.4.1
Monitoring to be done by discharger	3.3.2
Enforcement, including monitoring and surveillance by or on behalf of the control authority	3.1.6, 3.3.1.4
Ambient air quality standards	3.2.1, 3.2.4, 3.3.2
Rights of the individual	12, 3.4.1.2

* References are to section numbers.

237

[i] I.e. controls over products introduced for the purpose of protecting the external environment.

The Law and Practice Relating to Pollution Control in Luxembourg

Regulations Relating to Environmental Protection

1 CONSTITUTION

Articles 36, 37 const.

(ability of Grand-Duke and Ministers to make decrees and regulations)

2 CIVIL CODE

Article 544 and
Articles 1382 and 1383

(on the rights of property owners)
(responsibilities of operators of dangerous, dirty or noxious establishments)

3 PENAL CODE

Article 552

(on punishments for deposition of injurious objects or unhealthy emissions on the public highway)

Articles 561 and 562

(concerning punishments for those found guilty of causing a noise or disturbance at night)

4 CODE OF CRIMINAL INSTRUCTION

Article 119

(concerning objections to administrative actions)

5 SPECIAL LAWS

(List of principal laws and decrees referred to in the text.) (Abbreviations: GDD = Grand-Ducal Decree; GDR = Grand-Ducal Regulation; MD = Ministerial Decree.)

GDD 14 Dec 1789 GDD 16–24 Aug 1790 Law 27 June 1906	(power of communes to issue regulations safeguarding human health)
Law 27 Nov 1980	(created the Administration of the Environment)
Law 27 July 1979	(created the Higher Council for Nature Conservation)
Law 16 April 1979	(control of dangerous, dirty and noxious installations)
GDR 16 April 1979	(determining the list and classification of dangerous, dirty and noxious installations)
GDD 17 June 1879	(on decisions of the Burgomaster relating to dangerous, dirty and noxious installations)
Law 29 July 1965 Law 7 Sept 1978	(on participation in national affairs of associations of national importance)
Law 21 June 1976	(on atmospheric pollution)
Law 23 March 1963	(protection of the population against dangers arising from ionising radiation)
GDR 8 May 1981	(appointing experts and agents responsible for investigation and verification of regulations concerning pollution)
GDR 18 May 1979	(requirements for oil central heating installations)
GDR 12 July 1978	(sulphur content of liquid fuels)
GDR 20 July 1977	(lead content of petrol)
GDR 21 March 1980	(approval of motor vehicles)
GDR 28 Aug 1924	(health and safety of employees of industrial and commercial companies)
Law 20 March 1974	(land-use planning)
Law 21 June 1976	(noise control)
GDR 13 Feb 1979	(restrictions on noise levels outside establishments)
GDR 16 Nov 1978	(noise limits for music inside establishments)
GDD 23 Nov 1955 amended by GDD's of 11 April 1964, 13 May 1966 and 26 July 1980	(provisions of the Highway Code controlling noise from road traffic)
GDR 25 May 1979	(on approval of vehicles or parts of vehicles)
GDD 15 Sept 1939	(controlling the use of radiophonic equipment gramophones and loudspeakers)

Law 27 Aug 1927 (amended)	(forbidding noisiness in the area of schools or religious buildings whilst in use)
Ordinance of 13 Aug 1669	(prohibiting blockage of navigable rivers by building or deposition of rubbish)
Law 16 May 1929	(cleaning, maintenance and improvement of watercourses)
MD 9 Sept 1929	(requiring communes to present plans for water purification)
Law 21 March 1947 (amended)	(prohibiting discharge into watercourses of matter likely to harm fish)
Law 9 Jan 1961	(protection of watercourses)
GDD 17 Aug 1963	(appointing government delegates to oversee implementation of legislation on groundwaters)
GDR 17 May 1979	(on bathing waters)
GDR 20 Dec 1980	(on freshwater quality)
GDR 13 Nov 1970	(water intended for human consumption)
GDR 27 Aug 1977	(surface waters for drinking)
Law 17 July 1962	(protecting the supply of drinking water from Esch-sur-Sûre reservoir)
GDR 8 July 1963	(constituting the Syndicate for waters from the Esch-sur-Sûre reservoir)
GDR 14 Sept 1963	(procedure for inquiries into works at Esch-sur-Sûre dam)
Law 27 May 1961	(establishing protected zone around Esch-sur-Sûre reservoir)
GDR 21 March 1980	(determining activities prohibited in part 2 of zone around Esch-sur-Sûre reservoir)
Law 26 June 1980	(disposal of waste)
Law 27 July 1978	(prohibiting dumping of wastes except in designated locations)
Law 20 March 1974	(giving powers to the Government to make development plans)
GDR 26 June 1980	(on disposal of waste oils)
GDR 26 June 1980	(on disposal of waste from TiO_2 industry)
GDR 26 June 1980	(on disposal of PCBs and PCTs)
Law 25 March 1963	(outline law on protection of the population from dangers of radiation)
GDD 8 Feb 1967	(controlling establishments importing, distributing and transporting radioactive substances)
Law 25 Sept 1953	(foodstuffs, drinks and commodities)
GDR 8 June 1977	(positive list of preservatives for use in human foods)
GDR 9 Oct 1979	(emulsifying, stabilising, thickening and gelling agents for use in human foods)
GDR 25 Feb 1980	(control of meats and certain foods)
GDR 21 Dec 1980	(special foods)

GDR 13 April 1978	(materials and objects in contact with foodstuffs)
GDR 26 Nov 1979	(use of additives in animal foods)
GDR 24 Oct 1978	(ingredients and additives used in cosmetics)
Law 15 April 1980	(use of certain detergents in washing powders)
Law 28 Feb 1968	(pesticides and agricultural pharmaceuticals)
GDR 29 May 1970	(approval system for pesticides and agricultural pharmaceuticals)
Law 26 Feb 1973	(controls over fertilizers)
GDR 24 Jan 1979	(enforcement of the law on fertilizers)
Law 19 Feb 1973	(medicinal substances and the fight against drug addiction)
Law 14 March 1979	(classification, packaging and labelling of dangerous substances)
Law 17 July 1978	(amending Law 29 July 1965 on the conservation of nature and natural resources)
Law 12 June 1937	(planning of towns and built-up areas)
Law 20 March 1974	(land-use planning)

6 COURT DECISIONS

Superior Court of Justice

10 Nov 1971	(responsibility of successive polluters)
18 Jan 1964	(drought not constituting *force majeure,* nor torrential rain a 'state of necessity')
26 June 1979	(right of landowners to water above and below their land—limited interpretation of abuse of rights)

7 INTERNATIONAL CONVENTIONS

The Convention of 27 November 1886 between the Grand-Duchy and Belgium concerning the control of watercourses bordering the two countries.
The Franco-Belgian-Luxembourg Protocol of 8 April 1950 creating a permanent tripartite commission for polluted waters.
The Protocol of 20 December 1961 between the Federal Republic of Germany, France and the Grand-Duchy creating an international commission for the protection of the Moselle against pollution.
The Convention of 27 October 1956 between the same states for the canalisation of the Moselle.

The Berne Agreement of 29 April 1963 concerning the international commission for the protection of the Rhine against pollution.

The European Agreement on the limitation of the use of certain detergents in washing and cleaning products, signed in Strasbourg on 16 September 1968.

The Treaty of 17 October 1974 between the Grand-Duchy and the Land of Rhine Palatinate concerning the common realisation of tasks regarding water conservation by communes and other corporations (e.g. treatment plant on the Sûre at Echternach).

The Convention relating to the protection of the Rhine against pollution by chlorides.

The Convention relating to the protection of the Rhine against chemical pollution, and

The additional Agreement to the Agreement, signed in Berne on 29 April 1963, concerning the international Commission for the protection of the Rhine against pollution, signed in Bonn on 3 December 1976.

The Convention of 17 March 1980 between the Grand-Duchy and Belgium concerning the waters of the Sûre (construction of a treatment plant).

Decision M(78)10 of 14 November 1978, and decision M(79)2 of 4 May 1979, of the committee of Ministers of the Benelux Economic Union on sampling and analysis of fertilizers.

Convention on civil liability in the field of nuclear energy, Paris, 29 July 1960.

Convention relating to the liability of the operator of a nuclear installation, Brussels, 31 January 1963.

8 DIRECTIVES OF THE COUNCIL OF THE EUROPEAN COMMUNITIES

78/665/EEC 14 July 1978	(air pollution from motor vehicle exhausts)
70/157/EEC 6 Feb 1970	(permitted sounds from motor vehicle exhausts)
73/350/EEC 7 Nov 1973 ⎱ 77/212/EEC ⎰	(amending directive 70/157/EEC)
70/338/EEC	(acoustic warning devices)
74/151/EEC 4 March 1974	(noise level of agricultural vehicles)
77/311/EEC 29 March 1977	(cab noise level of agricultural vehicles)
76/160/EEC 8 Dec 1975	(bathing water quality)
78/659/EEC 18 July 1978	(quality of freshwater)
75/440/EEC 16 June 1975	(surface waters for drinking)
75/439/EEC 16 June 1975	(disposal of used oils)
78/176/EEC 20 Feb 1978	(disposal of TiO_2 waste)
76/403/EEC 6 April 1976	(disposal of PCBs and PCTs)
77/94/EEC 21 Dec 1976	(foodstuffs for particular nutritional uses)
76/893/EEC 23 Nov 1976	(materials and objects in contact with foods)
77/535/EEC 22 June 1977	(control of fertilizers)
79/138/EEC 14 Dec 1978	(control of fertilizers)

1
Political and Administrative Institutions

1.1 CENTRAL ORGANISATIONS

The country of Luxembourg is the largest of the small European states, although it represents only 0.2% of the surface area of Europe (2,586 square kilometres). It is a representative democracy in the form of a constitutional monarchy.

Unlike the Belgian Constitution (Articles 25–31), the Constitution of the Grand-Duchy contains no formal scheme for division of power; it arises from the layout and context of the Constitution.[1]

Legislative power is exercised jointly by the Chamber of Deputies and the Grand-Duke. The latter can present proposals or draft laws to the Chamber ; he also ratifies the laws (Article 34, Constitution). The Council of State has a right of suspensive veto in legislative matters.[2]

According to Article 33 of the Constitution, *executive power* is exercised by the Grand-Duke only, but all governmental acts issued by the Grand-Duke must be countersigned by a competent minister.

> In fact, the Grand-Duke is supported in the exercising of executive power by his Government which acts in his name, assumes the responsibility for his acts and in this way, participates in his representative character.[3]

1.2 JUDICIAL POWER

Judicial power is exercised by the courts and tribunals. As in Belgium, the responsibility for judging disputes relating to subjective rights rests with the ordinary tribunals, but the legislator can remove from them responsibility for judging disputes relating to political rights. Luxembourg is divided into twelve judicial cantons, each with its own justice of the peace, and into two districts, Luxembourg and Diekirch, each with a district tribunal. The Superior Court of Justice located in Luxembourg acts as Court of Appeal to judge appeals against judgements pronounced by the district tribunals on civil, commercial and criminal matters.

The Council of State is the supreme court on matters of administrative contention; it sometimes functions as an appeal authority and sometimes as an administrative court.[4]

1.3 LOCAL ORGANISATIONS

The territory of the Grand-Duchy is divided into 3 administrative districts, at the head of which there is a district commissioner appointed by the Grand-Duke; into 12 cantons which have no administrative role; and into 126 communes. The commune is the only example of the principle of territorial decentralisation. Communal autonomy is demonstrated by the fact that the members of the communal council are elected by the electorate of the commune for a term of 6 years. The daily administration of the commune is the responsibility of the College of Burgomaster and Aldermen consisting of one Burgomaster and two Aldermen.[5]

The Burgomaster is appointed by the Grand-Duke, generally from the members of communal council. The Aldermen for towns are appointed by the Grand-Duke and those for other communes by the Minister of the Interior; they are always chosen from the members of the communal council.

The district commissioners have a permanent supervisory role over the communes;[6] moreover, communal law has elaborated a system of central control of local government.

1.4 SOURCES OF LAW

Laws are the first expression of the will of the nation.

The Grand-Duke issues the regulations and decrees required to implement the laws, without ever being able to suspend the laws themselves or to dispense with their implementation (Article 36, Constitution).

A Grand-Ducal decree is always an implementing measure of a law. . . . The public administration regulation or *Grand-Ducal regulation* is a Grand-Ducal decree controlling matters of administration or general policing. . . . *The ministerial decree or regulation* is an implementing decree at a lower level.[7]

Article 37 of the Constitution gives the Grand-Duke the right to make *international treaties,*[8] but they have no effect until they have been approved by law and published in the form laid down for the publication of laws.

International institutions can be temporarily empowered, by treaty, to exercise the attributed powers reserved under the Constitution for the legislative, executive and judiciary authorities.

The Decrees of 14 December 1789 and of 16–24 August 1790, as well as the Law of 27 June 1906, enable the communes to issue *communal regulations* to safeguard public health.

The *orders and judgements* of the courts and tribunals themselves have a relative importance limited to the parties involved in the case; their legal force, once invested with the authority of a judgement, is valid only for the particular case in question and the decision given, with the exception of cancellation orders issued by the Council of State which are binding on all. In addition, no judge is bound to follow previous decisions; there is no doctrine of precedent.

Nevertheless, case law does, in fact, possess a certain power of persuasion; it also plays an important role in the formation of law.

1.5 ADMINISTRATIVE INSTITUTIONS AND THE CONTROL OF POLLUTION

While several ministerial departments have been made responsible for controlling certain aspects of pollution, the major role is undoubtedly played by the *Ministry of the Environment,* and especially by the *Administration de l'Environnement,* which was created by the Law of 27 November 1980.

The Administration of the Environment is responsible for the protection of the environment 'with a view to a better quality of life for mankind in its surrounding', namely by:

prevention of pollution and nuisances;

improvement of the fundamental conditions for environmental health by the fight against water and air pollution, noise control and waste disposal;

promoting environmental health with a view to safeguarding a well-balanced ecosystem;

studying and evaluating the impact on the environment of industrial, agricultural and urban activities;

executing laboratory programmes concerning the environment, at the demand of public authorities, enterprises and private persons;

research concerning the environment;

supervision and control of compliance with legal and regulatory provisions concerning the environment;

participation in the elaboration of these regulations;

collaboration with the other State and communal agencies, associations of communes, public corporations and international institutions dealing with problems concerning environmental quality;

information and promotion of any effort with a view to the protection of the environment.

The Administration of the Environment consists of a directorate and three services.

The *Water Service* is responsible for:

ensuring the safeguarding and management of water resources;

drawing up the inventory of discharges into surface waters;

drawing up the inventory of the quality of surface and ground waters and monitoring of changes;

elaborating the national quality assurance programme and ensuring its execution;

control of all drinking water operating facilities as well as of all facilities concerning disposal and purification of waste-waters;

carrying out analyses and tests concerning the quality of surface and ground waters, bathing waters and drinking waters;

working out appropriate analytical techniques and undertaking, on behalf of the other services, special laboratory research, which is not covered by their own monitoring networks.

The *Air and Noise Service* is responsible for:

ensuring the safeguard of the atmospheric environment by appropriate measures with a view to preventing air pollution and noise nuisances;

drawing up an inventory of emissions into the atmospheric environment and control of their generation;

drawing up an inventory of air quality and of noise levels;

promoting the creation of protected zones, for which specific measures are to be taken;

control of purification installations;

carrying out analyses and tests concerning air quality and noise levels.

The *Waste Service* is responsible for:

ensuring waste management by appropriate measures in order to promote prevention, re-use and treatment of waste;

drawing up an inventory of household refuse and of industrial, toxic and dangerous wastes, and of their impact on the environment; permanent monitoring of their generation;

promoting the execution of the national waste disposal plan dealing with the selection, collection, transport, treatment and storage of waste;

control of discharges and all other waste treatment plants;

carrying out analyses and tests concerning the composition of waste.

The *General Commission for Water Protection* is in charge of the coordination and direction of all measures to be taken for the protection of watercourses.

Other services responsible for the control of pollution are:

Ministry	Responsibilities
Ministry of State	Town and country planning: general policy and coordination
	Commission for national sites and monuments
Ministry of Foreign Affairs	European Communities
Ministry of Agriculture	Water and forests
	Nature conservation

Ministry of Economy and the Middle Classes	Industrial and food supplies
Ministry of Energy	General energy policy, supply and pricing
	Solid, liquid and gaseous fuels
	Production and distribution of energy
	Operation of State power stations
	Energy conservation and rational use of energy; alternative energy
	Storage of petroleum products
Ministry of the Interior	Syndicates of communes
	Superior Council for Water Distribution
	Urgan development; communal planning service
Ministry of Transport, Communications and Information	Road traffic
	Registration and technical control of vehicles
Ministry of Labour and Social Security	Inspection of Labour and Mines
Ministry of Public Works	Navigable and floatable watercourses
	Construction and maintenance of purification plants, sewers and pipes

The present Minister of the Environment is at the same time Minister of Transport, Communications and Information and Minister of Energy.[10]

1.6 CONSULTATIVE BODIES

There are various consultative bodies in the Grand-Duchy of Luxembourg. The most important is undoubtedly the Higher Council for

Nature Conservation which, under the terms of Article 18 of the coordinated laws concerning the protection of the natural environment, is responsible for:

1   presenting proposals on its own initiative to the Government on nature conservation matters;

2   giving opinions on all questions and drafts submitted to it by the Government;

3   giving opinions on all the measures to be taken in implementation of the present law.

The Council consists of 8 members appointed by the Minister for a term of 4 years, presided over by the director of the Water and Forests Administration. The mandate of outgoing members is renewable. In the event of a vacancy, the Minister appoints a new member who is in office for the remainder of the term of mandate of his predecessor.

The explanatory text giving the reasoning behind the draft law which became the Law of 27 July 1978 contains the following statements:

> Article 18. The problems of the natural and human environment assume an ever-growing importance. In consequence, it appeared logical to entrust the Higher Council for Nature Conservation with tasks which emphasise its essential role as coordinator and catalyst on these matters and also its duties of submitting to the Government proposals relating to policy on environmental matters.
>
> For this reason, as well as to enable the administrations and services (Bridges and Roads, Agricultural Technical Services) involved to be represented on the Council, the number of Council members has been increased to eight.[11]
>
> Article 19. The Higher Council for Nature Conservation will be called upon to exercise important functions in the context of town and country planning with regard to the sectoral plan for the protection of the natural environment. It has therefore proved necessary to define the new powers attributed to this Council.
>
> For this reason, and to enable the other ministerial departments (Tourism, Agriculture, Public Works) to be involved in the Council, the number of Council members has been increased. This is in order to give representation to the various ministries and not to the different administrations and services as indicated in the comments on the articles of the initial draft law (Doc. Parl. No. 1729, ordinary session, 1972–1973).[12]

The existence of the following organisations should also be noted:

The Higher Council for Land-Use Planning,[13] consisting of a president and 15 members made up as follows:

(a) 2 representatives of communes, delegates of the Association of Luxembourg Towns and Communes;

(b) 3 district commissioners;

(c) 2 delegates of the Economic and Social Council;

(d) 1 architect, delegate of the Order of Architects;

(e) 1 delegate from private environmental protection or nature conservation organisations;

(f) 1 representative of the Farmers' Union;

(g) 1 representative of the National Council of Trade Unions;

(h) 1 representative of the Federation of Industries;

(i) 3 individuals appointed on personal merit.

The secretary responsible for management of the land-use planning secretariat, appointed by the Ministry for Land-Use Planning, is by right a member of the Council.

The Technical Council for Water Sanitation,[14] under the presidency of the Commissioner for Water Protection, to which is entrusted the elaboration of advice concerning the general planning and coordination of the measures to be taken in the medium- and long-term, as well as necessary immediate measures.

1.7 NATIONAL COMMITTEE FOR THE PROTECTION OF THE ENVIRONMENT

This Committee was created on 29 June 1981. The Committee is responsible for:

1 carrying out, or ordering to be carried out, environmental impact assessments;

2 advising on, and coordinating opinions on each application for authorisation submitted in conformity with laws and regulations concerning complementary aspects of environmental protection;

3 editing and supplementing the permanent code 'Environmental Management and Protection'.

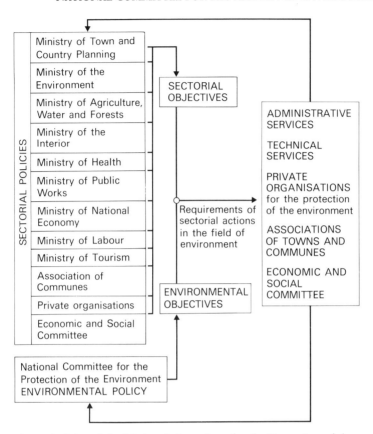

Operation principles of the National Committee for the Protection of the Environment.

1.8 NON-GOVERNMENTAL ORGANISATIONS

The association, the Luxembourg League for the Protection of Nature and Environment (NATURA) has as its aims:

(a) The defence of the idea of stronger protection of nature and the environment, of a healthy conservation of natural environment and of a more rational use of natural resources;

(b) To draw the attention of the public at large and of the relevant authorities to the necessity to protect nature and the environment, especially the typical sites, the fauna and flora of the country, to inform the public on the means to achieve this protection, including the fight against pollution of soil, water, and the atmosphere;

(c) to bring together, with a view to dynamic action, the non-official organisations which are active in the field of nature and environment protection.

The means of action used by NATURA in order to realise its aims include, amongst others:

petitions addressed to the public authorities; all other means of interventions with the authorities;

conferences and seminars, either public or private, information in the press and on radio and TV; publication of a periodical.

The number of effective members, associations for the protection of nature and environment, is now 33.

Notes

1. See Pierre Majerus, *L'état luxembourgeois*, 3rd edition, 1970, p. 114; F. Welter, Speech delivered on the occasion of the Council of State Centenary, in *Conseil d'Etat. Livre jubilaire*, 1957, pp. 49–51.
2. Under the terms of Article 59 of the Constitution, all the laws are subject to a second vote unless the Chamber decides otherwise, in agreement with the Council of State. If the Council of State refuses to dispense with the second vote, there is an interval of at least three months between the two votes by the Chamber. The Council of State therefore exercises the moderating influence of a second legislative assembly; compare the House of Lords in England and the 'Eerste Kamer' in the Netherlands.
3. Pierre Majerus, *op. cit.*, p. 128.
4. See the organic law of 8 February 1961.
5. Except in the town of luxembourg where there are six aldermen.
6. With the exception of Luxembourg Town which remains under the direct supervision of the Minister of the Interior and the Government.
7. P. Majerus, *op. cit.*, p. 128.
8. With the exception of commercial treaties and tariff and payment agreements in the name of the Belgo-Luxembourg Economic Union; as a result of Article 31 of the coordinated Convention instituting UEBL, Belgium concludes these treaties, reserving the facility for the Luxembourg Government to sign them jointly with the Belgian Government.
9. *Mémorial*, 1981.
10. Grand-Ducal decree of 23 July 1979 attributing ministerial departments to members of the Government.
11. *Doc. Parl.*, Chamber of Deputies, 1972–1973, No. 1720.
12. *Doc. Parl.*, Chamber of Deputies, 1977–1978, No. 1729–2.
13. Grand-Ducal regulation of 27 September 1976, *Mémorial*, 28 December 1976.
14. Ministerial regulation of 3 June 1965, *Mémorial*, 1965, p. 97.
15. Address: Place d'Armes, Luxembourg. Tel. 25588.

2
The Control of Dangerous, Dirty or Noxious Installations

2.1 INTRODUCTION

The Luxembourg legislation relating to dangerous, dirty or noxious installations is very similar to the Belgian legislation on the subject.

The regulations which currently apply are the Law of 16 April 1979 relating to dangerous, dirty or noxious installations and the Grand-Ducal Regulation of 16 April 1979 determining the list and classification of dangerous, dirty or noxious installations.[1] As is the case in Belgium, the Luxembourg system is based on three fundamental principles:

(a) the necessity of obtaining authorisation from the administrative authority before establishing an installation included on the list, or transferring it from one location to another;

(b) permanent monitoring of authorised installations;

(c) the safeguarding of the rights of third parties.

Since the administration must take account of the question of protecting industrial and professional freedom,[2] an important innovation contained in the new Law of 1979 consists of the fact that henceforth installation authorisations can be refused not only for reasons of safety, cleanliness or convenience in relation to the public, to the neighbourhood or to the workforce, but also when the nature of the establishment, installation or process is such that it may damage the environment.

261

2.2 FIELD OF APPLICATION[3]

The Law of 16 April 1979 applies to 'all public or private industrial, commercial or workshop establishments', as well as 'all manufacturing installations or processes', if they are mentioned in the list attached to the Grand-Ducal Regulation of 16 April 1979. The dangerous, dirty or noxious character of the existence or operation of these establishments, installations or processes—referred to henceforth as establishments—is therefore the object of an irrefutable legal presumption. The establishments are divided into three classes, instead of two classes as is the case in Belgium.

The existence, operation, transfer, extension or conversion of establishments in classes 1 and 2 is subject to authorisation.

A new authorisation is necessary:

(1) if the establishment did not start operating within the period stipulated in the authorisation decree;

(2) if it has been out of action for 2 consecutive years;

(3) if it was destroyed or temporarily momentarily put out of action by an accident of any kind resulting from its operation.

Before they can open or start operating, class 3 establishments are the subject of a written declaration to the Inspection of Works and Mines. They are subject to general requirements dictated by the Grand-Ducal Regulation.

In principle, the authorisation is attached to the property and passes to a new owner in the event of the property changing hands.[4]

2.3 THE OBLIGATION TO OBTAIN AN AUTHORISATION

2.3.1 The responsible authorities

Establishments in class 1 are authorised by the Minister of Labour, after consulting the Minister responsible for the protection of the environment; establishments in class 2 are authorised by the Burgomaster.

When the establishment to be erected or transferred includes several kinds of operation, one of which comes under class 1, a decision on the entire establishment is taken by the Minister of Labour.

2.3.2 Preliminary enquiry—organisations consulted

Authorisation applications for class 1 establishments are sent to the Inspectorate of Labour and Mines. Authorisation applications for class 2 establishments are sent to the Burgomaster of the commune in which the proposed establishment is to be located. Article 6 of the Law of 16 April 1979 determines the information to be included in the application; in particular, it must contain the measures planned to prevent or reduce any inconvenience which may result from the establishment, with regard to the workforce, the neighbourhood, the public and the environment.

An important innovation consists of the ability of the authority responsible for granting the authorisation to require a summary assessment of the eventual environment impact of the proposed establishment.

In this context, it must also be pointed out that the Administration of the Environment contacts the promoters of new industries before any decision to start the official authorisation application procedures is made.

In this way, it is possible, by a first summary assessment of the environmental impact, to point out clearly to the industrialist the conditions of operation which will be imposed.[5]

Authorisation applications are posted by the College of Burgomaster and Aldermen for a period of 15 days in the commune in which the establishment is to be located, as well as in neighbouring communes within a radius of 200 m from the establishment.

Notices must be simultaneously posted at the town hall and, in a very visible position, at the site of the proposed establishment; from the date when these notices appear, the application and plans are placed on view and can be consulted by all interested parties at the town hall of the commune in which the proposed establishment is to be located.

In addition, in locations with more than 10,000 inhabitants, authorisation applications for class 1 and class 2 establishments are brought to the attention of the public by the publication of extracts in at least four daily newspapers printed and published in the Grand-Duchy. The same applies to class 1 establishments in other localities. The publication costs are paid by the applicant.

When the period of 15 days has expired, a member of the College of Burgomaster and Aldermen, or a commissaire specially designated for the purpose by the Burgomaster, collects the written observations and proceeds with a public hearing in the commune in which the establishment is to be located, at which any interested party may be heard. A written report on the enquiry is then prepared. For class 1 establishments the file, together with proof of publication of the application notice, the written report on the enquiry and the opinion of the College of Burgomasters and Aldermen, are forwarded to the Inspectorate of Labour and Mines.

The publication of the application and of the *de commodo et incommodo* enquiry constitute essential formalities, the omission of which leads to the cancellation of the authorisation decision.[6]

Even though the Law of 16 April 1979 does not explicitly mention it, the responsible authority can also request the opinion of other administrations concerned with the application.

2.3.3 Decision

The Law does not dictate the period within which the responsible authority must reach its decision.

Under the Grand-Ducal Decree of 17 June 1872, it was judged that, as with all administrative acts and in the absence of a legal instruction on the subject, the decisions of the Minister or the Burgomaster were not required to indicate the reasoning on which they were based, provided that such reasoning existed and conformed to the Grand-Ducal Decree.[7]

2.3.4 Responsibility for assessment

The authorisation fixes the reservations and conditions of operation which are considered necessary in the interests of safety, cleanliness or convenience, as well as in the interest of the environment.

Under the Grand-Ducal Decree of 1872, the Council of State specified that the following factors could not be taken into consideration:

the fact that the neighbourhood surrounding a dangerous, dirty or noxious establishment would be subject to a sharp decrease in property value;[8]

town-planning and aesthetic considerations;[9]

economic considerations such as the damage resulting to competitors;[10]

considerations arising from the interests of tourism.[11]

No condition can be imposed, the application of which 'would, by itself, render illusory, if not impossible, the proposed operation'.[12]

2.3.5 Duration of validity

Authorisations can be of limited duration; there is, however, no legal limit.

The authorisation fixes the period within which the establishment must commence operating. As already mentioned, if this period is exceeded, a new authorisation application must be made.

2.3.6 Publicity for the decision

Applicants are notified through administrative channels[13] of decisions to grant, refuse or withdraw authorisation; decisions granting authorisation are immediately posted up by the communal authorities in all the communes involved (the Law of 16 April 1979 does not state for how long the notice should be displayed).[14]

All parties who presented written observations to the enquiry are informed by registered letter that a decision granting, refusing or withdrawing authorisation has been taken.

A copy of every decision of the Burgomaster concerning a class 2 establishment is sent to the Minister of Labour and the Inspectorate of Labour and Mines.

2.3.7 Appeal to the Council of State Committee of Contention

Article 13 of the Law of 16 April 1979 grants the right of appeal to the authorisation applicant and to all interested parties in general. The

265

Council of State, Committee of Contention (*Comité Contentieux*), acts in the last instance as fundamental judge.

The 'interest' must arise from injury to a right of maintaining the previous situation.

Since Article 25 of the Law of 29 July 1965, added to by the Law of 7 September 1978, admits that associations of national importance approved by the Minister of the Interior can be called upon to participate in the action of public authorisations the purpose of which is the protection of the environment, it would appear that an appeal by such an association should be acceptable.

An appeal must be lodged, at the risk of forfeiture, within 30 days.

This appeal period commences from notification (for the applicant) or from the date when the decision notice is posted up (for other interested parties), or from the date when the decision was brought to the attention of the Government (for the Minister of Labour, in the event of a decision by the Burgomaster relating to a class 2 establishment).

The appeal has no suspensive effect; however, the Council of State can grant a suspension of execution if there are urgent and exceptional circumstances.[16]

The neighbours of a dangerous, dirty or noxious establishment can intervene in the procedure by which the operator requests a review of decision to refuse an authorisation.[17]

2.4 MONITORING OF THE APPLICATION OF THE AUTHORISATION

2.4.1 Monitoring

The Burgomaster is in charge of the permanent monitoring of authorised establishments; the 'overall monitoring' is carried out under the care of the Inspectorate of Labour and Mines. These authorities ensure that the conditions imposed on the operator are observed.

2.4.2 Sanctions

2.4.2.1 ADMINISTRATIVE SANCTIONS

New conditions can be imposed at any time; the authority which granted the authorisation can revoke it. The operator cannot make use of an acquired right resulting from the fact that the authorisation has been in existence for a long time.[18]

The authority which granted the authorisation can at any time ensure that the conditions imposed on the operator of an establishment under the present law are observed. The authorisation can be withdrawn by a justified decision by the authority which granted it, if the candidate does not observe these decisions, or if he refuses to submit to the new obligations which the responsible authority always has the right to impose.

The operator has the right of appeal to the Committee of Contention of the Council of State. Such an appeal must be lodged within 30 days from notification.

If an establishment reputed to be dangerous, dirty or noxious operates without authorisation, in spite of authorisation having been refused, or if the conditions imposed are not observed, the director of the Inspectorate of Labour and Mines for classes 1 and 3, and the Burgomaster of the commune for class 2, can cause the cessation of operations by provisional measures and, if necessary, have the establishment closed down and the equipment sealed off.

An appeal can be made to the Committee of Contention of the Council of State against any decision to suspend, close down or seal off an establishment; this appeal must be lodged without fail within 10 days.

All interested parties, for example the residents of the area, can at any time make complaints to the Burgomaster or the Inspectorate of Labour and Mines against any inconvenience caused by the operation of an establishment. In fact, the number of complaints is very small; the annual average since 1965 amounts to about 50 complaints. The number of contraventions ascertained and dealt with is even less significant: on average about 5 per year.[19]

2.4.2.2 PENAL SANCTIONS

Contraventions of the requirements of the Law of 16 April 1979 and of its implementing regulations and decrees are punishable by imprisonment from 8 days to 6 months and/or a fine of 2,501–1,000,000 Luxembourg Francs. The starting up or operation of an establishment considered to be dangerous, dirty or noxious without authorisation constitutes a continous contravention; public action only ceases on the day when the operation ceases.

2.4.2.3 CIVIL SANCTIONS

The operating authorisation has no detrimental effect on the rights of third parties.

The responsibility of the operator is that of common law. The control of establishments considered to be dangerous, dirty or noxious does not deprive an action based on Articles 1382 and 1383 of the Civil Code or Article 544 of the Civil Code of its rightful character, nor does it remove it from the jurisdiction of the ordinary tribunals.[20]

While, because of the principle of the division of powers, it is not up to the judiciary authority to prescribe any measure which is the exclusive responsibility of the administrative authority, the right to order reparation, that is to say the restitution of the situation as it was prior to the contravention, implies the right to order the closure of an establishment which was installed and set in operation without authorisation from the administration.

In ordering this measure, the tribunals do not encroach upon the responsibilities of the administrative authority and do not censure an administrative act, but make a ruling within the framework of their own responsibilities, on an application from a civil party for damages or reparation.[21]

Notes

1. *Mémorial*, 27 April 1979, p. 678–693. The Grand-Ducal Decree of 17 June 1872, amended by that of 7 July 1882, and the Grand-Ducal Decree of 1 August 1913 revising the list of establishments are repealed. The Grand-Ducal Decree of 17 June 1972 repealed the Royal Decree of 31 January 1824.
2. Council of State, 27 March 1974, *Pas.* XXII, 475.
3. In addition to the general regulation, there is a specific regulation relating to the transport and sale of explosive materials (Law of 20 April 1881, *Mémorial*, 1881, p. 281) and a special law concerning the protection of the population against the dangers resulting from ionising radiation (*Mémorial*, 1963, p. 227).
4. It was judged that this is no longer the case when the authorisation is granted to the lessee of a property in order to establish a temporary factory and the owner-lessor has stipulated the removal of the factory when the lease expires. In this case, the effect of the authorisation, which is acquired by the factory and not by the rented property, disappears with the establishment without the landlord, lessor or his assigns being able to oppose the renunciation of the authorised manufacture or the modification of the authorisation by the Government (Council of State, 9 February 1882; *Pas.* II, 74; 21 June 1972, *Pas.* XXII, 184).
5. Minister of the Environment, *Exposé budgétaire,* November 1981, p. 17.
6. Council of State, 20 July 1978, *Pas.* XXIV, 91; 9 March 1982, not published.
7. Council of State, 28 November 1935, Nemers case, not published; 22 May 1935, Pas. XIII, 427. See, however, the Law of 1 December 1978 controlling the non-contentious administrative procedure, *Mémorial*, 1978, p. 2486.

8. Council of State, 14 July 1948, Anton and cons. case, not published; comp. trib. Luxembourg, 28 February 1894, *Pas.* III, 456; Luxembourg, 1 May 1912, *Pas.* VIII, 413.
9. Council of State, 13 March 1964, *Pas.* XIX, 261.
10. Council of State, 15 May 1918, Glesener case, not published.
11. Council of State, 21 January 1970, *Pas.* XXI, 253.
12. Council of State, 21 May 1980, *Pas,* 1981, 9.
13. Since the notification is not assumed, it is up to the authority to prove that it has carried out the necessary formalities for the appeal period to be valid; Council of State, 30 June 1951, *Pas,* XV, 146.
14. The Council of State has stated: 'The posting up of the notice having the aim and the effect of giving rise to the very brief period for contentious appeal, it must be done in such a way as to inform all those people who may have an interest in attacking the decision.' (Council of State, 26 January 1973, *Pas.* XXII, 263).
15. Compare Alex Bonn, *Le contentieux administratifen droit luxembourgeois,* 1966, p. 135, no. 140 and the cases mentioned therein.
16. Council of State, 14 July 1948, in the Anton case, not published; in Belgium the Council of State is not empowered to pronounce a suspension of execution.
17. Council of State, 12 November 1975, *Pas.* XXIII, 215.
18. Council of State, 12 November 1975, *Pas.* XXIII, 215.
19. See the annual reports of the Inspectorate of Labour and Mines. In 1975 there were 53 complaints and three contraventions were ascertained and dealt with.
20. Superior Court, 15 December 1911, *Pas.* VII, 518.
21. Superior Court, 17 January 1957, *Pas.* XVII, 106.

3

Atmospheric Pollution

3.1 THE LAW OF 21 JUNE 1976 RELATING TO THE FIGHT AGAINST ATMOSPHERIC POLLUTION[1]

3.1.1

The outline Law of 21 June 1976 covers widely differing sources of pollution. In fact, in the context of this Law, atmospheric pollution is taken to mean

> all emissions into the air, *whatever the source,* of gaseous, liquid or solid substances, in quantities and concentrations liable to cause an abnormal nuisance to man or to result in damage to his health, to injure animals or plants or to cause damage to goods or sites.[2]

The Law does not, however, apply to pollution of the atmosphere resulting from ionising radiation; this is controlled by the Law of 23 March 1963 concerning the protection of the population against dangers arising from ionising radiations.[3]

3.1.2

The Law enables the Grand-Duke to fix all measures to be taken to prevent, reduce or suppress pollution of the atmosphere.

These Grand-Ducal regulations can only be issued after obligatory consultation of the Council of State and the consent of the Commission of Labour of the Chamber of Deputies. These regulations can:

(1) determine the cases and conditions in which the emission of gaseous, liquid or solid substances into the atmosphere is forbidden;

(2) control or prohibit all circumstances or activities generally liable to lead to atmospheric pollution, and in particular, the starting up, operation or use of certain industrial, commercial, craft or agricultural establishments, domestic central heating installations, equipment or devices and motor vehicles;

(3) impose and control the siting and use of equipment or devices intended to prevent or combat pollution;

(4) create protected zones and issue specific measures to be observed in these zones;

(5) organise a system of periodic checks to be carried out on central heating combustion installations and fix the cost of these checks, which is borne by the users of the heating installations.

The member of the government responsible for the environment is in charge of coordinating the actions of the authorities in this sector.

3.1.3 Supervising the implementation of the Law

Under Article 3 of the Law, the Grand-Ducal Regulation of 8 May 1980 appoints experts and agents who are responsible for investigating and verifying infringements of the legal and regulatory requirements concerning air pollution and noise.[5] These experts and agents can proceed with the control of all circumstances and activities liable to result in prohibited atmospheric pollution; in particular, in the presence of duly summoned interested parties, they can measure the emission of substances into the atmosphere. The individuals concerned are entitled to the assistance of an expert of their choice, but this may not delay the action of the official experts and agents. These can also test, or have tested, any equipment or device liable to cause pollution or intended to prevent pollution.

The operator of an establishment, in addition to his employees, the owners or lessees of a private dwelling, the owners and users of a motor vehicle and anyone responsible for a condition or activity generally presumed to be the cause of prohibited atmospheric pollution, is obliged, at the request of the agents, to assist them in carrying out their duties under the Law. Every owner or user of a motor vehicle is obliged to make that vehicle available to the agents during the period required for the test.

In the event of a conviction, the costs incurred by the measures taken under this Article are the responsibility of the owners, operators or users. In all other cases, the costs are borne by the State.

3.1.4

Grand-Ducal regulations will specify the powers of the agents, fix the methods and conditions whereby measurements and tests are to be carried out and will lay down all other necessary control measures for the implementation of the Law. These regulations have not yet been issued.

Nevertheless, the Law itself gives the experts and agents the power to enter all establishments, day or night, if they have reason to believe that the Law or regulations concerning the fight against atmospheric pollution are being infringed, with the exception of residential premises.

If there are sufficient indications to suppose that the cause of atmospheric pollution is to be found inside residential premises, a house visit can be made between 07.00 and 21.00 by two agents who have a warrant from the examining magistrate.

3.1.5 Imminent danger

In the event of imminent danger of prohibited atmospheric pollution, the member of government in charge of protecting the human environment can take the necessary emergency measures to deal with the situation and, in particular, forbid any activity liable to result in such pollution.

The decision must be notified by registered letter to the individuals affected by such measures. During the month of notification an appeal can be made to the Council of State Committee of Contention, which will take a final decision on the substance of the case.

3.1.6 Emergency measures

When a prohibited act of atmospheric pollution is committed, the examining magistrate can, at the request of the State Procurator or the civil party, order the necessary emergency measures to deal with the situation.

He can prohibit any activity which has resulted in pollution and forbid the use of and have sealed off, any equipment or devices which, because of their construction or characteristics, are not functioning in accordance with the implementing regulations of the Law.

The State Procurator, the polluter and the civil party can object to the orders of the examining magistrate.

The objection is brought before the Chamber of Indictment *(Chambre des mises en accusation)*. It is made, investigated and judged in accordance with the provisions of Article 119 of the Code of Criminal Instruction.

The General State Procurator also has the right to object. He must notify his objection within 10 days from the order of the examining magistrate.

The order is provisionally implemented.

3.1.7 Sanctions

Without prejudice to punishments provided for by other legal provisions, infringements of the Law of 21 June 1976 and its implementing regulations are punished by imprisonment from 8 days to 6 months and a fine of from 2,501–200,000 Luxembourg Francs, or by one of these punishments alone.

In the event of a repeated offence within 2 years, these punishments can be doubled.

3.2 OIL CENTRAL HEATING INSTALLATIONS

3.2.1

The Grand-Ducal Regulation of 18 May 1979 concerning the requirements for oil central heating installations and the control of these installations,[6] regulates the starting up and operating of oil central heating installations equipped with pulverising burners. This regulation does not apply to industrial, craft or agricultural installations which are submitted to a similar control under other legal or regulatory provisions and, in particular, on the subject of dangerous, dirty or noxious establishments.[7]

Oil central heating installations must be installed and operated in such a way that the smoke emitted by the chimney is less dark than scale 2 on the Ringelmann scale.[8] In addition, they must fulfil the following requirements:

(1) the soot index determined according to the method described in annex II of the regulation must not exceed the value 3;

(2) the combustion of gases must be such that there is no trace of oil or incompletely burnt oil particles in the soot deposit left on the filter, handled according to annex II;

(3) in installations brought into use before the present regulation came into force, the carbon dioxide (CO_2) content of the combustion gases must be at least 7%; for installations brought into use or substantially modified since the present regulation came into force, the CO_2 content of the combustion gases must be at least 10%;

(4) the temperature of the combustion gases must not exceed 300°C at the point where the soot index is measured.

3.2.2

Before a new oil central heating installation can be brought into use, or after it has undergone substantial alterations, it must be approved, on application from the person responsible for its installation, by one of the agents or experts appointed under Article 3 of the Law of 21 June 1976 relating to the fight against atmospheric pollution. The user is subsequently required to have the installation inspected and serviced every 2 years.

The first installation inspection was carried out when the 1979 regulation came into force and installations which were brought into use before the end of 1979 had to be inspected before 1 January 1981.

The inspection of central heating installations can be carried out by central heating installations enterprises or by legally established inspection companies. Inspections may be carried out only by a controller who holds a master's diploma or certificate established by the Chamber of Trade, endorsed by the Minister of the Environment, stating that he has achieved at least the level of a professional aptitude certificate in the occupation in question or in an associated activity, and that at the end of his training he has acquired the special recognition required, according to the state of the art provided for under the Grand-Ducal Decree of 18 May 1979.

If, during the inspection, it is found that an installation is not functioning in accordance with the requirements of the regulation, the controllers immediately proceed with the required adjustments.

If, nevertheless, they are not able to satisfy the requirements, they send a note to the Institute of Hygiene and Public Health informing them of the impossibility of adjusting the installation and the probable cause.

If the situation can be remedied by a simple maintenance operation, the user is given 1 month during which this must be carried out. If the installation requires modification, the user must carry this out within 6 months from the inspection. In both cases, a new inspection must be carried out within the specified period to ensure that the installation conforms with the requirements of Article 3 and can be kept in use.

If, during the inspection, the controller carrying out the inspection concludes that the installation requires modification, or, during the supplementary inspection, that the installation still does not conform with the requirements, the user can refer to another controller for a second opinion. In the event of a disagreement between the two controllers, a decision is taken by one of the agents or experts appointed by the Minister.

3.2.3 Sanctions

Infringements of the provisions of this regulation are sanctioned by the punishments envisaged by the Law of 21 June 1976; see 3.1.7 above.

The following are considered to be infringements:

(1) the omission by an installer to apply for approval for a plant he has installed;

(2) the starting up of an installation by the user before approval has been obtained;

(3) failure on the part of the user to have an inspection carried out;

(4) the operation by the user of an installation which does not comply with the requirements after the permitted period has elapsed;

(5) the refusal by a user to present the results of the last inspection to the controlling agents.

3.3 THE SULPHUR CONTENT OF CERTAIN LIQUID FUELS

The Grand-Ducal Regulation of 12 July 1978 concerning the sulphur content of certain liquid fuels,[9] which applies to oil[10] used as fuel but excludes oil used in power stations, states that after 1 October 1980 it is forbidden to deliver, use, import, store or transport for sale oils with a sulphur compound content expressed as sulphur exceeding 0.3% in weight (0.5% for light fuel oil).

There is, however, an exception to this regulation in Article 7:

> If, because of a sudden change in crude oil supplies, alterations occur to the sulphur content of this petrol, which, taking into account the lack of available desulphurisation capacity, may compromise supplies to consumers, the Government Council can admit into the territory of the Grand-Duchy and authorise the storage and transport for sale, the delivery and use of fuel oils which do not conform to the specifications under Article 3. The Council immediately informs the Commission of the European Communities of its decision, and within three months and after consultation with the other Member States, the Commission decides on the methods and duration of this derogation.

A decision taken by the Government Council on 25 May 1979 made use of this provision. Since the Grand-Duchy is dependent on the Belgian and Dutch refineries, it was 'obliged for the time being to take supplies of crude oil rich in sulphur'. During the period from 26 August 1979 to 1 October 1980 the storage and transport for sale, delivery and use of fuel oils with a sulphur compound content expressed as sulphur of up to 0.8% in weight was authorised.

3.4 POLLUTION ARISING FROM MOTOR CARS

3.4.1

The Grand-Ducal Regulation of 20 June 1977[12] prohibits the importation, sale, storage or transport for sale and use of petrol with a content of lead compounds expressed as lead exceeding 0.55 g/l.

At present (since 1 January 1981), the content of lead compounds, calculated as lead, must not exceed 0.40 g/l.

This regulation specifies that the achievement of a reduction in the lead content does not authorise an alteration in the composition of the petrol liable to increase excessively the quantities of other pollutants contained in the exhaust fumes.

3.4.2

The Grand-Ducal Regulation of 21 March 1980[14] states that vehicles and tractors must be approved in accordance with the provisions of the Commission of the European Communities Directive 78/665/EEC of 14 July 1978, updating Directive 70/220/EEC of the Council on the legislation of Member States concerning measures to be taken against air pollution from the exhaust fumes of combustion engines in motor vehicles.

3.4.3

Under the terms of Article 25, paragraph 3, of the Highway Code, motor vehicles and motor cycles must not emit fumes which can damage traffic or inconvenience other road users.

3.5 INDUSTRIAL POLLUTION

The control of classified establishments applies to all establishments contained in the list attached to the Grand-Ducal Regulation of 16 April 1979.[16] The authorisation can lay down conditions to prevent or combat atmospheric pollution.

In addition to these provisions on the subject, there is also the Grand-Ducal Decree of 28 August 1924 relating to the health and safety of personnel employed by industrial and commercial companies.

3.6 OTHER LEGISLATION

3.6.1

As a result of the first article of the coordinated Laws concerning the protection of the environment,[17] no building can be commenced or erected outside a built-up area without authorisation from the Minister responsible for the administration of water and forests.

Article 1, paragraph 2, prohibits the authorisation of a project constituting a danger for conservation (especially conservation of the atmosphere).

3.6.2

It goes without saying that development plans can have an effect on the protection of the atmosphere, particularly through provisions relating to the siting of industrial zones. The Law of 20 March 1974 concerning land-use planning in general states that development plans should contribute in particular to the improvement of living conditions of the population and to the cleanliness of the environment (Article 2).

3.6.3 Penal Code

Article 552, paragraph 1 of the Penal Code punishes with a fine of 50–200 Luxembourg Francs those who throw, leave or abandon on the public highway objects which can injure as a result of falling or of their unhealthy exhalations.

3.6.4 Communal regulations

The Decrees of 14 December 1789 and 16–24 August 1790, as well as the Law of 27 June 1906, on the protection of public health, authorise the communes to issue police regulations or to take certain sanitary measures to enable the inhabitants to enjoy cleanliness in the streets, public spaces and buildings.

Notes

1. *Mémorial*, 1 July 1976.
2. Compare the definition of the Belgian Law of 28 December 1964, under Belgium, Chapter 3, paragraph 3.1.1.
3. See Chapter 7.
4. Without prejudice to the powers of the judiciary police officers, the officers of the gendarmerie and the police.

5. *Mémorial*, 23 September 1977.
6. *Mémorial*, 4 July 1979.
7. See Chapter 2.
8. See annex I of the Grand-Ducal Decree of 18 May 1979.
9. *Mémorial*, 22 August 1978.
10. 'Fuel oil' is taken to mean all petroleum products which, because of their distillation, are one of the medium distillates intended for use as fuel, of which at least 85% in volume, including distillation losses, are distilled at 350°C.
11. *Mémorial*, 28 June 1979.
12. *Mémorial*, 6 July 1977.
13. 'Petrol' is taken to mean all fuel used for internal combustion engines with automatic ignition.
14. *Mémorial*, 28 March 1980.
15. *Official Journal,* 14 August 1978, No. L223.
16. See Chapter 2.
17. *Mémorial*, 3 October 1978.
18. *Mémorial*, 23 March 1974.

4

Noise Control

4.1 THE LAW OF 21 JUNE 1976 RELATING TO NOISE CONTROL[1]

4.1.1 Definition of noise

The outline Law of 21 June 1976 defines noise as follows:

acoustic emissions which, whatever their source, result in damage to health, the capacity to work or the well-being of man. (Article 1)

4.1.2 Scope of the regulations

The Law enables the Grand-Duke to fix all the measures to be taken to prevent, reduce or suppress noise.

Grand-Ducal Regulations can be issued only after seeking the opinion of the Council of State and with the assent of the Commission of Labour of the Chamber of Deputies.

In particular, these regulations can:

(1) prohibit the production of certain noises;

(2) make the production of certain noises subject to restrictions, among others, to restrict the times when the noise can be produced;

(3) control or prohibit the manufacture, import, export, transit, transport, offer for sale, sale, free or conditional gift, distribution, installation and use of apparatus, devices or objects which produce, or are liable to produce certain noises;

(4) to impose and control the siting and use of apparatus or devices intended to reduce, absorb or remedy noise and its inconveniences;

(5) to create protected zones and decree specific measures which must be observed in these zones;

(6) to impose technical conditions regarding construction and installation likely to attenuate the inconveniences of noise and its spread.

The member of government responsible for the environment is in charge of coordinating action by the authorities concerning noise control.

4.1.3 Supervision of the application of the Law

Infringements of the Law and its implementing regulations are investigated and verified by officers of the judicial police, agents of the gendarmerie and the local police, as well as the experts and agents appointed under the Grand-Ducal Regulation of 8 May 1981.[2]

They can check on any state or activity generally likely to cause noise; in the presence of interested parties or those duly summoned by them, they can test or have tested any apparatus or device likely to produce noise, as well as those intended to reduce or absorb it, or remedy its inconveniences. In the event of conviction the costs of the tests are borne by the owner. In all other cases, the costs are borne by the State.

They have the power to enter any establishment, night or day, if they have reason to believe that an infringement of the law or regulations relating to noise control is being committed, with the exception of residential premises. If there is sufficient indication to suppose that the cause of a noise is to be found inside residential premises, a house visit can be made between 07.00 and 21.00 by two agents acting on a warrant issued by the examining magistrate.

4.1.4 Imminent danger

In the event of imminent danger of prohibited acoustic emissions, the member of government responsible for protection of the human environment can take necessary emergency measures to deal with the situation and, in particular, can prohibit any activity liable to produce emissions of that kind.

The decision is notified by registered letter to the individuals affected by the measure. During the month of notification an appeal can be made to the Council of State, Judicial Committee, which will take the final decision on the substance of the case.

4.1.5 Emergency measures

When prohibited acoustic emissions have taken place, the examining magistrate, at the request of the State Procurator or the civil party, can order whatever emergency measures that the situation requires. In particular, he can prohibit any activity which has caused these emissions and forbid the use and seal off equipment or devices which, because of their construction or characteristics, are not in a fit state to function in accordance with the implementing regulations of the Law.

The State Procurator, the person causing the prohibited emissions and the civil party can object to the orders issued by the examining magistrate. The objection must be lodged with the Chamber of Indictment. It is lodged, examined and judged in accordance with the provisions of Article 119 of the Code of Criminal Instruction.

The State Procurator General also has the right to object. He must lodge his objection within 10 days following the order of the examining magistrate.

The order is provisionally implemented.

4.1.6 Sanctions

The sanctions envisaged are identical to those under the Law of the same date relating to atmospheric pollution.[3]

The Law expressly states that it does not detract from the powers of the communal authorities to take 'all measures intended to guarantee public peace and quiet', particularly as a result of the Decrees of 14 December 1789 and 16–24 August 1790.

4.2 IMPLEMENTING REGULATIONS OF THE LAW OF 21 JUNE 1976

4.2.1 Establishments and construction sites

The Grand-Ducal Regulation of 13 February 1979[4] controls the noise level of the immediate surroundings of establishments—that is to say, 'all industrial, craft, commercial, agricultural or horticultural businesses, whether private or public'—and construction sites—that is to say, 'all sites used for construction, land development, repairs, earthworks or warehousing, whether public or private'.

Immediate surroundings are taken to mean the area within the boundaries with neighbouring property in which people are permanently located, in whatever capacity, either frequently or at regular intervals.

4.2.2 Limits

The regulation recommends those responsible for establishments and construction sites in built-up areas[5] not to exceed the following noise levels in the immediate surroundings, depending on the nature of the area:

Zone	Noise level (dB (A)) day	night	Nature of the area
I	45	35	Hospitals, recreation areas
II	50	35	Rural area, peaceful habitat, little traffic
III	55	40	Urban area, majority of residences little traffic
IV	60	45	Urban area with a few factories or businesses, medium traffic
V	65	50	Town centre (businesses, offices, entertainments), heavy traffic
VI	70	60	Predominance of heavy industry

Outside built-up areas the recommended levels are those indicated for zone VI; however, if the noise emitted by these establishments and construction sites can be perceived inside a built-up area, the recommended level measured at the edge of the built-up area is that indicated for the zone in question.

For construction sites, the fixed levels can be exceeded by:

20 dB (A) if the works last less than 1 month;

15 dB (A) if the works last for between 1 and 6 months;

10 dB (A) if the works last for between 6 months and 1 year.

4.2.3 Construction work at night

It is forbidden to exceed the recommended sound levels in a permanent or regular manner by more than 10 dB (A).

Inside built-up areas construction work is forbidden during the night. Under special circumstances and by prior request before the works commence, the Minister responsible for the Inspectorate of Labour and Mines can waive this prohibition with the approval of the Institute of Hygiene and Public Health. In this case, the maximum permissible night-time sound levels apply. Unless the ministerial authorisation decree contains indications to the contrary, the increases in sound level envisaged for construction sites are not applicable.

4.2.4 Transitional provisions

The Grand-Ducal Regulation of 13 February 1979 does not waive any more severe conditions which may have been imposed by the controlling authority under the legal provisions relating to classified establishments.[7]

The Minister of the Environment can grant a 3 year dispensation from observing the criteria to establishments which were already in existence when the regulation came into force, i.e. 21 June 1979. This dispensation is considered granted if, 1 month after the application is presented, no negative decision has been notified to the applicant.

In exceptional circumstances the Minister can grant a dispensation for a period of up to 10 years in the case of an establishment of particular economic importance for the town or region in which it is located, and if the observation of the fixed criteria is not technically possible, or if it would require alterations which could seriously compromise the establishment's ability to compete.

However, during these transition periods, establishments with a dispensation can under no circumstances emit a noise level which exceeds the level they were emitting when the regulation came into force. All new services created or equipment installed after the regulation came into force must conform to the prescribed requirements.

Finally, if a zone changes in nature in such a way as to make the noise criteria more severe for establishments in the area, these establishments are allowed a period of 3 years in which to comply with these new obligations. After this 3 year period there is still the possibility of an exceptional dispensation for 10 years.

4.2.5 Music inside establishments and in their surroundings

The Grand-Ducal Regulation of 16 November 1978[8] fixes the acoustic levels for music inside establishments and in their surroundings, that is to say, inside rooms and buildings located in the immediate surroundings in which people are to be found.

Music is taken to mean all methods of emitting electronically amplified music from the permanent or temporary sound source.[9]

As in Belgium, in public establishments[10] the maximum sound level emitted by music must not exceed 90 dB (A). This sound level is measured at any point in the establishment where people are normally present.

The sound level in the neighbourhood of music produced in a public establishment, or elsewhere, must not:

exceed the background noise by 5 dB (A) if it is less than 30 dB (A);

exceed 35 dB (A) when the background noise level is between 30 and 35 dB (A);

exceed the background noise level when it is over 35 dB (A).

'Background noise level' is taken to mean the minimum noise level measured over a period of 5 minutes, excluding 'music' in the meaning of this regulation.

The sound level is measured inside the room or building with the doors and windows closed. The microphone is placed at least 1 m away from the walls and at a height of 1.2 m above floor level.

The sound level in dB (A) is measured with the help of a sound meter which must conform with the requirements of the International Electro-technical Commission's recommendations:

IEC No. 123: recommendations relating to sound meters;

IEC No. 179: precision sound meters.

In addition, the sound meter must be regulated on an 'A' moderation filter and 'rapid measure'.

Before each measurement or series of measurements relating to the same sound source, the sound meter must be adjusted with the help of a standard acoustic source.

It should be noted that the Minister responsible for the environment is authorised to grant waivers to the levels envisaged, but only 'in exceptional cases and for limited periods'.

4.3 ROAD TRAFFIC NOISE

Several provisions in the Highway Code—Grand-Ducal Decree of 23 November 1955, amended by Grand-Ducal Decrees of 11 April 1964, 13 May 1966 and 26 July 1980—concern the control of noise from road traffic.

Under the terms of Article 25 *ter*, motor vehicles and motor bikes must not 'produce noise liable to cause inconvenience to users of the adjoining area'. They must be fitted with an exhaust device which must be adequately silenced.

The use of a motor vehicle which produces excessive noise is prohibited.

Articles 131–133 of the Highway Code contain the following provisions:

the use of acoustic warning devices for purposes other than safety is forbidden;

acoustic warning devices should never be used in an exaggerated fashion;

in built-up areas it is forbidden at all times to use an acoustic warning device except in the case of imminent danger;

outside built-up areas, acoustic warning devices cannot be used between dusk and dawn except in the case of imminent danger;

between dawn and dusk the acoustic warning device should be used outside built-up areas:

before overtaking another road user;

on the approach to places where visibility is insufficient;

every time it is required for the purposes of road safety.

The Grand-Ducal Regulation of 25 May 1979 states that the approval of vehicles or parts of vehicles as well as tractors or parts of tractors can only be carried out in accordance with the provisions of the following Council of the European Communities' Directives:

70/157/EEC of 6 February 1970 (permissible sound level and exhaust devices for motor vehicles) *Official Journal (OJ)* L42, 23 February 1970, amended by Directives 73/350/EEC of 7 November 1973, *OJ* L321, 22 November 1973 and 77/212/EEC of 8 March 1977, *OJ* L66, March 1977;

70/388/EEC of 27 July 1970 (acoustic warning device for motor vehicles) *OJ* L176, 10 August 1970;

74/151/EEC of 4 March 1974 (sound level of agricultural or forestry tractors on wheels) *OJ* L84, 28 March 1974;

77/311/EEC of 29 March 1977 (sound level for drivers of agricultural or forestry tractors on wheels) *OJ* L105, 28 April 1977.

4.4 OTHER LEGISLATION

4.4.1

Article 561, paragraph 1, and Article 562 of the Penal Code punish with a fine of 200–400 Luxembourg Francs and imprisonment from 1–5 days, or by one of these punishments alone, those who are guilty of causing a noise or disturbance at night which disrupts the peace and quiet of the inhabitants. In the event of a repeated offence, the punishment can be increased by an extra 9 days' imprisonment.

4.4.2

A Grand-Ducal Decree of 15 September 1939 controls the use of radiophonic equipment, gramophones and loudspeakers. It is forbidden during the day to operate radiophonic equipment or gramophones in such a way that they disturb the peace and quiet of the public by the excessive intensity or power of broadcasting equipment. At night, equipment of this kind can be operated only at a very muted level.

The use of loudspeakers installed outside the house or broadcasting their sound outside, as well as mobile loudspeakers, is prohibited. The Minister of the Interor can lift this prohibition for specific cases.

Infringements of these requirements are punishable by imprisonment from 1 to 7 days and a fine of 200–500 Luxembourg Francs, or by one of these punishments only.

Confiscation of the equipment can be ordered.

4.4.3

Finally, there is a somewhat folkloristic provision on the control of taverns (Article 23 of the amended Law of 12 August 1927):

> It is forbidden to go to public meetings and to hold noisy parties in the neighbourhood of building used for religious purposes during divine service, as well as during lesson time in the vicinity of buildings in which teaching takes place, on pain of a fine of 6–30 Luxembourg Francs to be paid by each offender.

The fine will be 12–600 Luxembourg Francs for all those who continue the disturbance after receiving an injunction from the prosecuting officers.

Notes

1. *Mémorial,* 1 July 1976.
2. See 3.1.3.
3. See 3.1.7.
4. *Mémorial,* 21 March 1979.
5. 'Built-up area' is taken to mean all groups of at least 5 houses which, during at least 3 months of the year, are used for human habitation, and located within a radius of 100 m. For the application of this provision to establishments with the exception of construction sites, a property which, though not yet built, is liable to be converted as a result of a building permit under the existing communal regulations, is considered to be a property in which people reside.
6. The period between 07.00 and 22.00 is considered as 'day' and that between 22.00 and 07.00 hours as 'night'.
7. See Chapter 2.
8. *Mémorial,* 14 December 1978.
9. Compare the Belgian definition under Belgium, Chapter 4, para. 4.2.2.2.
10. 'Public establishments' means all establishments as well as their annexes which are

accessible to the public, even if their access is limited to certain categories of persons, by payment or not, such as dance halls, concert halls, discotheques, private clubs, shops, restaurants, bars, including those located outdoors.

5
Water Pollution

5.1 INTRODUCTION

5.1.1

Intervention by the Luxembourg legislator on the subject of water pollution has been considerable; some such activities date back to early days.

In this context it should be noted that the ordinance on Water and Forests of 13 August 1669, Article 42, Heading XXVII stated that

> no one, whether owner or tenant, can construct mills, embankments, locks, weirs, sluices or walls, plant trees, heap up stones, earth or sticks, or construct other buildings or obstructions harmful to the flow of water in navigable rivers, nor throw any refuse or filth, or accumulate such matter on piers or river banks, on pain of arbitrary fine.[1]

Of course, more recent legal texts are more specific and appropriate. It is no longer the punitive approach which prevails, but rather a policy of prevention.

5.1.2

Since the creation of the *General Commission for Water Protection* in 1963, a serious treatment programme has been carried out. This Commission plays a role of direction and coordination. It is responsible for the technical and financial conception and planning of water policy as a whole. To this end, it has the necessary finance available from annual payments into the 'Special Fund for the Purification of Watercourses', currently standing at about 400 million Luxembourg Francs (370 million Luxembourg Francs in 1982).

5.2 THE LAW OF 16 MAY 1929 CONCERNING CLEANSING, MAINTENANCE AND IMPROVEMENT OF WATERCOURSES[2]

5.2.1 Introduction

The Law of 16 May 1929 controls water pollution. Article 13 states:

> It is forbidden to throw, pour or allow to flow into the watercourses, either directly or indirectly, any matter likely to harm the conservation of the water, the free flow of the water, its cleanliness, its use for animal watering, domestic, agricultural or industrial purposes, or for irrigation, the breeding or conservation of edible fish or crustaceans or the culture or conservation of aquatic flora which can be used for any purpose whatever.

However, waste water from industry or communal built-up areas can be discharged directly or indirectly into watercourses, provided it has undergone an effective pretreatment from an organoleptical, physical, chemical and bacteriological viewpoint. Waste-water may only be discharged into the ground by communes or individuals provided it does not jeopardise the use of groundwaters and is not harmful to health.[3]

5.2.2 Control of industrial pollution

Article 14 of the Law of 16 May 1929 states:

> A ministerial decree will fix for each industry . . . the conditions under which residues will be discharged.

Under the scheme of the Law, an industry which intends to discharge waste-water into a watercourse must build a treatment plant, have this plant approved by ministerial decree and, comply with special conditions imposed by the decree concerning the method of discharge if it is to avoid the general prohibition on discharges.

It is considered that the ministerial decree provided for in Article 14 of the Law of 16 May 1929 is not a general measure to be issued spontaneously by the Government, but an individual provision to be issued in each case at the request of the interested party, and that the term 'industry' does not mean different types of industrial activity, but industrial establishments in operation.[4]

5.2.3 Communal built-up areas

The Ministerial Decree of 9 September 1929 required communes which discharge their waters into a watercourse without pretreatment, or adequate pretreatment, to present their treatment plans within '6 months'. Under Article 14 of the Law, these must be approved by ministerial decree.

The total number of purification plants in operation at the end of 1980 is shown in the following table:

Plant type	Plant capacity (no. of inhabitants)					Total
	≤500	≤1000	≤3000	≤10,000	≤100,000	
Mechanical plants equipped with lagooning	205	10	2			217
Mechanical plants	3					3
Biological plants:						
with bacterial filter	5	2		1		8
with activated sludge	11	7	13	6	10*	47
Total	224	19	15	7	10	275

*The Beggen plant has a capacity equivalent to 300,000 inhabitants[5]

The total number of biological plants increased to 55 at the end of 1980, with a total treatment capacity equivalent to 756,650 inhabitants.

5.2.4 Appeal

An appeal can be made against decisions of the Government under Article 14 of the Law, to the Council of State Committee of Contention, which has direct jurisdiction.

This appeal must be made by the communal administration or individual concerned within 1 month from notification of the decision.

5.2.5 Sanctions

Punishment by a fine of 2000–30,000 Luxembourg Francs will be imposed on

> Whoever allows to flow, throws or deposits into a watercourse liquids or matter likely to taint or alter the water, particularly by the discharge of waste-water arising from industrial establishments or local drainage, without having obtained authorisation as envisaged under Article 14 or in infringement of the conditions imposed by the ministerial decree. (Article 17)[6]

In the event of the prescribed works, orders or judgements issued under the law not being carried out, they can be implemented by the administrative authority at the expense of those responsible for the infringement. These costs are recovered by a simple statement of account, in the same way as revenue taxes are collected.

The penal provisions of the Law of 16 May 1929 apply even to involuntary acts of pollution. Infringement is committed by the material violation of the legal provision in question, in the absence of any intentional element.

The polluter cannot be exonerated unless he succeeds in establishing in his favour a state of necessity, *force majeure* or an act by a third party.[7]

It has been considered that a season of abnormal drought cannot constitute a *force majeure* for the polluter; equally, torrential rain in the area cannot constitute a state of necessity.[8]

5.2.6 Monitoring the application of the Law

Agents of the public works administration and the agricultural service, duly sworn in, as well as all judiciary police officers have the right to verify and lodge information concerning infringements relating to watercourses.

5.2.7 Fishing controls

To complete the picture, the prohibition contained in Article 42, paragraph 2 of the amended Law of 21 March 1947 concerning fishing controls in national waters should be mentioned:

> it is forbidden to discharge matter and residues into watercourses likely to harm fish or destroy the aquatic fauna and flora, arising from factories or other industrial or private establishment.

Article 45, paragraph 8, of this Law provides the following sanctions: a fine of 5000–75,000 Luxembourg Francs and/or imprisonment from 8 days to 6 months.

Contrary to the penal provisions contained in the Law of 16 May 1929, Article 45, paragraph 8, covers only intentional acts of pollution.

5.3 THE LAW OF 9 JANUARY 1961 ON THE PROTECTION OF GROUNDWATERS

5.3.1

As a result of the Law of 9 January 1961,[9] all abstractions of groundwater and associated installations are subject to prior authorisation from the Minister of the Interior. The expression 'abstraction of groundwater' includes all emergency water collection points, wells, boreholes, galleries, drainage and all works and installations in general for the purpose of abstracting groundwater.

However, all groundwater offtakes of a depth equal to or less than 20 m in a non-gushing water table and operated manually, as well as drainage installations and operations to reduce the water table which do not lead to its reduction to more than 2 m below the level of the natural terrain, are exempt from authorisation.

On the other hand, the following are considered to be equivalent to the establishment of a new groundwater offtake:

> all alterations which would result in an existing water offtake ceasing to comply with the exemption conditions mentioned above;

the extension or alteration to a groundwater offtake previously exempt from authorisation;

the resumption of use of groundwater offtakes which have been out of service for a continuous period of 5 years.

Finally, the commercial development and operation of mines in the sandstone of Luxembourg are subject to prior authorisation from the Ministers of the Interior, Public Health and Public Works.

5.3.2

Authorisation must be applied for by the owner, the contractor undertaking the works and the operator. Authorisations can be withdrawn or suspended if the conditions to which they are subject are not observed. New conditions can always be imposed.

Reasons for ministerial decisions must be given. They can be referred to the Council of State Committee of Contention, which decides the substance of the case.

5.3.3

It is forbidden for the owner to discharge into the ground any matter which is likely to contaminate groundwaters.

5.3.4 Monitoring the application of the Law

A Grand-Ducal Decree of 17 August 1963[10] appoints 'government delegates' in charge of monitoring the implementation of the provisions of the Law of 9 January 1961 and its implementing decrees:

(a) from the bridges and highways administration: the geological service engineer and the foremen inspectors;

(b) from the water and forestry administration: the engineer inspectors;

(c) from the public health administration: the medical inspectors;

(d) from the inspectorate of labour and mines: the director or his delegate;

(e) from the agricultural services administration: the foremen inspectors.

In carrying out their duties these delegates have the same powers as judicial police officers. They verify infringements and submit reports which are taken as fact in the absence of proof to the contrary. Their responsibility extends throughout the territory of the Grand-Duchy.

5.3.5 Sanctions

Infringements of the provisions of the Law and its implementing regulations are punished by a fine of 1000–30,000 Luxembourg Francs. The judgement can order confiscation of machines and the demolition of works and also the restoration of land to its original state.

Any heads of businesses, owners, operators, employers, directors, managers or others in charge who have obstructed government delegates in carrying out their monitoring duties are liable to the fine specified.

Heads of businesses, owners, users and operators are legally responsible for fines imposed on their directors, managers or others in charge.

5.4 BATHING WATERS

The Grand-Ducal Regulation of 17 May 1979 concerning the quality of bathing waters[11] ensures the inclusion in internal law of the Council of the European Communities Directive 76/160/EEC of 8 December 1975.

During the bathing season (from 15 May to 31 August) bathing waters must conform to the physiochemical and microbiological parameters contained in the annex to the regulation. They are considered to conform to these parameters:

if samples of the water, collected from the same place, in accordance with the frequency laid down in the annex, show that 95% of the samples comply with the parameters fixed in the annex, except with regard to the 'total coliform' and 'fecal coliform' parameters where the percentage can be 80% of the samples; and

if for the respective 5% and 20% of samples which do not conform:

the water quality is not more than 50% above the parameter figure in question, with the exception of microbiological, PH and dissolved oxygen parameters, consecutive samples of water collected at a statistically appropriate frequency do not vary from the relevent parameter figures.[12]

The regulation sets out in detail how the sampling and consecutive analyses should be carried out. The Institute of Hygiene and Public Health or any other laboratory approved by the Minister of the Environment carries out inspections and sampling.

The Minister of the Environment, with the agreement of the Minister of Public Health, can make exemptions to the provisions of this regulation:

(a) for certain parameters marked (0) in the annex, for exceptional meterological and geographical reasons;

(b) when bathing waters are subject to a natural increase in certain substances which result in the limits fixed in the annex being exceeded.

In no case can exemptions to this regulation have a detrimental effect on public health protection requirements.

The Institute of Hygiene and Public Health regularly informs the operator of a bathing zone of the results of inspections and sampling operations which have been carried out either by the Institute or by an approved laboratory. If the water does not comply with any one of the levels fixed in the annex, the operator is ordered by registered letter to cease operations immediately and to erect visible notices in the area indicating that bathing is prohibited.

The operator of a bathing zone who fails to erect visible signs prohibiting bathing in the bathing zone, who fails to keep the signs visible as long as the bathing waters do not conform to the required standard, or who fails to cease operations as soon as he receives the registered letter ordering him to do so from the Institute of Hygiene and Public Health, is committing an infringement which is punishable by imprisonment from 8 days to 6 months and a fine of 250–50,000 Luxembourg Francs, or by one of these punishments alone.

If the bathing zone is not operated commercially, the Institute of Hygiene and Public Health will erect signs to prohibit bathing and also informs the Burgomaster of the commune and the State Procurator responsible for the area in question.

The results of analyses carried out in 1979 show that the chemical and bacteriological standards were met in the Upper Sûre,[13] the Our and the Wiltz; they were not achieved in the Lower Sûre.

5.5 FRESH WATERS NEEDING PROTECTION TO SUPPORT FISH LIFE

Implementing the EEC Directive 78/659 of 18 July on the quality of fresh waters needing protection or improvement in order to support fish life, a Grand-Ducal Regulation of 20 December 1980[14] aims at the protection or improvement of the quality of those running or standing fresh waters which support or which, if pollution were reduced or eliminated, would become capable of supporting fish belonging to:

indigenous species offering a natural diversity;

species, the presence of which is judged desirable for water management purposes.

The designated waters are divided into salmonid waters and cyprinid waters.[15] The physical and chemical parameters applicable to these waters are listed in an annex to the regulation.

In the case of non-compliance with the values set in the annex, the competent Ministers shall establish whether this situation is due to unforeseen factors or whether it is the result of a natural phenomenon or of pollution; they shall adopt appropriate measures with a view to controlling any eventual pollution.

The Minister of the Environment shall establish, if necessary, a treatment programme to ensure that the designated waters conform with the values of the annex within 5 years following their designation.

5.6 DRINKING WATER

5.6.1

The Grand-Ducal Regulation of 13 November 1970 relating to water intended directly or indirectly for human consumption[16] fixes the hygiene standards to which drinking water should conform, that is to say, the standards for some 20 organoleptical and chemico-physical parameters as well as 6 microbiological parameters. In fact, with a few rare exceptions, the drinking water supply of all communes conforms to these standards.

As a result of the EEC Directive of 15 July 1980 relating to the quality of water intended for human consumption, our Grand-Ducal

Regulation on this matter will be modified in the course of 1982. The new regulation will not only fix the maximum concentrations for numerous parameters—chemico-physical as well as microbiological parameters—but also indicate the minimum frequency of sampling, as a function of the population served.[17]

5.6.2

The Grand-Ducal Regulation of 27 August 1977 concerning the required quality of surface water intended for use as drinking water embodied EEC Directive 75/440/EEC of 16 June 1975 in Luxembourg internal legislation. The matter is now regulated by the Grand-Ducal Regulation of 12 June 1981.[18]

5.6.3

The Law of 17 July 1962 intended to reinforce the supply of drinking water in the Grand-Duchy of Luxembourg from the Esch-sur-Sûre reservoir[19] authorises the State, the Water Syndicate of the South, the Syndicate responsible for the operation and maintenance of the water network in the Ardennes and the town of Luxembourg to form an organisation for the establishment, maintenance and operation of all works, mechanical installations and drainage networks used for drinking water taken from the Esch-sur-Sûre reservoir. The 'Syndicate for Waters from the Esch-sur-Sûre Dam' was constituted by Grand-Ducal Decree on 8 July 1963.

The Syndicate is authorised to create installations to convey water from Esch-sur-Sûre to different parts of the country and to construct and operate a water treatment plant.

The enquiry procedure to be implemented before the works are carried out is specified by a Grand-Ducal Regulation of 14 September 1963.[20]

5.7 HEALTH PROTECTION FOR THE ESCH-SUR-SÛRE DAM

The reservoir above the Esch-sur-Sûre dam should be considered from two different aspects. Firstly it constitutes a health investment since it is one of the nation's drinking water resources which must be protected by all available means. Secondly, it is situated in very beautiful country-side and the reservoir and shores provide numerous and varied recreational possibilities. Every effective health protection policy for the waters of the reservoir must reconcile the different interests at stake, without leading to a loss in the socio-economic development of an entire region.[21]

To this end, the Law of 27 May 1961[22] established a protected zone around the Esch-sur-Sûre dam.

This health protection zone is made up of two parts. In Part 1 of the health protection zone it is forbidden:

(a) to construct residential houses, weekend cottages, garages, stables, barns, silos, workshops, industrial and commercial establishments;

(b) to develop boreholes, ditches, mines;

(c) to discharge and treat waste-waters and to deposit rubbish;

(d) to fish, swim, engage in nautical sports or use boats of any kind;

(e) to camp;

(f) to carry out any installation or activity whatsoever which might soil or disturb the waters of the lake.

These provisions do not apply to activities which are necessary for the monitoring and operation of the lake and the dam.

The Grand-Ducal Regulation of 21 March 1980[23] determines the installations, works and activities which are forbidden in Part 2, or which, without prejudice to the formalities required under other legal and regulatory provisions, are subject to prior authorisation from the Minister of Public Health.

In particular, it is forbidden to discharge untreated waste-waters and to discharge hydrocarbons, especially used oil. All new hydrocarbon storage installations for private use, the development of boreholes, wells, ditches and mines, the discharge of treated waste-waters, the installation and operation of bathing establishments and the spreading of manure and organic and mineral fertilizers, as well as the use of products to combat crop pests and diseases on a strip of land 100 m wide from the edge of the reservoir, are subject to special authorisation from the Minister of Health.

Infringements of the provisions of the Law of 27 May 1961 are punishable by imprisonment from 8 days to 1 year and a fine of 500–100,000 Luxembourg Francs, or by one of these punishments alone. Infringements of the regulations issued in implementation of the Law are punishable by imprisonment from 8 days to 3 months and a fine of 500–25,000 Luxembourg Francs, or by one of these punishments alone.

The court can officially order the demolition of constructions undertaken in infringement of these provisions, at the expense of the offender.

If infringements of the provisions of the Law or public administration regulations issued in implementation of the Law lead to pollution of water intended for public drinking use which results in the alteration of the health of an individual, they are punishable by imprisonment from 1 month to 2 years and a fine of 10,000–200,000 Luxembourg Francs, or by one of these punishments alone.

If infringements lead to the death of an individual, an apparently incurable illness, permanent incapacity for work or the absolute loss of an organ, they are punishable by imprisonment from 2 to 5 years and a fine of 200,000–500,000 Luxembourg Francs, or one of these punishments alone.

5.8 INTERNATIONAL CONVENTIONS

The Grand-Duchy of Luxembourg has signed and ratified various international conventions concerning water pollution.

The most important are:

(a) The Convention of 27 November 1886 between the Grand-Duchy and Belgium concerning the control of watercourses bordering the two countries;

(b) The Franco-Belgian-Luxembourg Protocol of 8 April 1950 creating a permanent tripartite commission for polluted waters;

(c) The Protocol of 20 Decmeber 1961 between the Federal Republic of Germany, France and the Grand-Duchy creating an international commission for the protection of the Moselle against pollution;

(d) The Convention of 27 October 1956 between the same states for the canalisation of the Moselle;

(e) The Berne Agreement of 29 April 1963 concerning the international commission for the protection of the Rhine against pollution;

(f) The European Agreement on the limitation of the use of certain detergents in washing and cleaning products, signed in Strasbourg on 16 September 1968;

(g) The Treaty of 17 October 1974 between the Grand-Duchy and the Land of Rhine Palatinate concerning the common realisation of tasks regarding water conservation by communes and other corporations (e.g. treatment plant on the Sûre at Echternach);

(h) The Convention relating to the protection of the Rhine against pollution by chlorides;

(i) The Convention relating to the protection of the Rhine against chemical pollution;

(j) The additional Agreement to the Agreement, signed in Berne on 29 April 1963, concerning the international Commission for the protection of the Rhine against pollution, signed in Bonn on 3 December 1976;

(k) The Convention of 17 March 1980 between the Grand-Duchy and Belgium concerning the waters of the Sûre (construction of a treatment plant).

Notes

1. 'Arbitrary fine' means 'an indeterminate fine which, in the present penal system, can only be expressed within the limits of the fine envisaged for the smallest first class police infringements, which is a fine from 1–10 Luxembourg Francs' (Cass. 30 July 1909 and 5 November 1909, *Pas. lux.*, VIII, 182).
2. *Mémorial*, 1929, p. 511. See also the ministerial decrees of 9 September 1929 and 22 December 1938 concerning the treatment of waste-waters arising from industry and drainage waters from communal built-up areas which are discharged into watercourses, *Mémorial*, 1929, p. 796 and 1938, p. 1354.
3. See 5.3.1–5.3.5.
4. Higher Court of Justice, 26 May 1951, *Pas. lux.*, XV, p. 154. In fact, model conditions limiting discharges into watercourses are repeated in the installation and operating authorisation granted as a result of the legislation on classified establishments (see Chapter 2). In addition, other conditions can be imposed in the context of a special permission called a 'watercourse' permission granted by the Minister of Agriculture, on the basis of the Law of 1669 on the Policing of Water and Forests, of the Decree of the *Directorire of 19 ventôse, year VI,* of the Royal Decree of 28 August 1920 and the Law of 16 May 1929.
5. Source: budgetary report of the Minister of the Environment, November 1980, p. 1.
6. As a result of Article 1, B, II, 19 of the Law of 26 February 1973, the court responsible is the police court.
7. R. Everling, 'The civil and penal responsibility of the polluter in Luxembourg law' in *Les aspects juridiques de l'environnement*, Namur, 1975, p. 124, and the jurisprudence cited therein.
8. Higher court of Justice, 23 December 1963 and 18 January 1964, unpublished decisions, cited by R. Everling, *op. cit.*, p. 124.
9. *Mémorial*, 1961, p. 17.
10. *Mémorial*, 1963, p. 355.
11. *Mémorial*, 4 July 1979, p. 1080.
12. Exceeding the limits is not taken into consideration in the discounting of percentages

when they are the result of flooding, natural catastrophes or exceptional meteorological conditions.

13. With the exception of Michelau (pollution by waste-waters from the camping sites at Moulin de Bourscheid).

14. *Mémorial*, 27 December 1980.

15. For the purpose of the regulation: 'salmonid waters' shall mean waters which support or become capable of supporting fish belonging to species such as salmon *(Salmo salar)*, trout *(Salmo trutta)*, grayling *(Thymallus thymallus)* and white-fish *(Coregonus)*; 'cyprinid waters' shall mean waters which support or become capable of supporting fish belonging to the cyprinids (Cyprinidae), or other species such as pike *(Esox lucius)*, perch *(Perca fluviatilis)* and eel *(Anguilla anguilla)*.

16. *Mémorial*, 3 December 1970, p. 1329. This regulation applies to water distributed by the public network to the population, to water used in the preparation or conservation of foods or beverages, including ice, and to water intended for use as conserved water delivered in bottles or other receptacles.

17. *Exposé budgétaire du Ministre de l'Environnement,* November 1981, p. 21.

18. *Mémorial*, 23 September 1977, p. 1570; *Mémorial*, 1981, p. 1066.

19. *Mémorial*, 28 August 1962, p. 898.

20. *Mémorial*, 30 September 1963, p. 807.

21. Extract of directives to be followed in the elaboration, revision or alteration of general development plans for communes situated in the basin of the Haute-Sûre lake, *Mémorial B*, 4 August 1978, p. 862.

22. *Mémorial*, 12 June 1961, p. 429.

23. *Mémorial*, 15 April 1980, p. 393.

6

The Disposal of Waste

6.1 THE SITUATION PRIOR TO THE LAW OF 26 JUNE 1980

6.1.1

Before the Law of 26 June 1980 on the disposal of waste came into force, waste control in Luxembourg centred around several texts, notably:

Article 4 of the Law of 27 July 1978 concerning the protection of the natural environment;[1]

the Law of 16 April 1979 relating to dangerous, dirty and noxious establishments;[2]

the legislation concerning land-use planning;[3]

the Law of 27 May 1961 concerning the sanitary protection measures for the Esch-sur-Sûre barrage.[4]

6.1.2

Article 4 of the Law of 27 July 1978 contains an absolute prohibition on abandoning, depositing or throwing of wastes of any kind, including obsolete mechanical engines and their parts, other than in specially designated locations selected for the purpose by the communal authorities.

The following are subject to authorisation by the Minister responsible for water and forestry administration:

(a) The installation and operation of a refuse tip. Authorisation is granted only if the tip (where waste is either to be buried or hidden from view) is managed in such a way as to avoid harmful or disagreeable emissions or dirty vapours.

(b) The deposition of waste, including all mechanical engines, by owners, tenants or users of the land.

(c) The deposition of industrial wastes and materials outside industrial zones shown in development plans provisionally approved and legally published by the communal councils (as long as the plans are not in the process of being opposed).

(d) The maintenance of deposits existing at the moment when the Law of 27 July 1978 came into force. The application for authorisation had to be made within 3 months of the publication of the law in *Mémorial*, i.e. 7 December 1978 at the latest.

The Law specifies that authorisation should be refused, after consultation with the member of the government responsible for the human environment, if the tip or deposit is liable to harm the beauty or character of the countryside or if it constitutes a danger for the conservation of land, the sub-soil, the atmosphere, the flora, fauna or natural environment in general.

An appeal can be made against the decisions of the Minister to the Council of State Committee of Contention, which decides in the last instance on the substance of the case.

6.1.3

The deposition of liquid and solid wastes, their treatment and incineration come under No. 201 of the classified establishment nomenclature and must always be authorised—as class 1 establishments—by the Minister of Labour, after consultation with the Minister of the Environment.

The same applies to deposits of chemical or toxic wastes and scrap metal. On the other hand, derelict car dumps are class II establishments which are authorised by the Burgomaster.

6.1.4

As a result of Article 10 of the Law of 20 March 1974 concerning the general development of the territory, the Government has decreed in council, in the form of a management programme, the priority objectives for the development policy of the territory and the measures to be applied with a view to its implementation.[5]

In addition, the Government can establish partial or global development plans.[6] A good example of this is the decision of the Government in Council of 16 June 1978 laying down directives to be followed during the elaboration, revision or modification of general development plans for the communes situated in the Haute-Sûre lake basin area.[7] These directives will serve as the basis for the elaboration of a global development plan.

With regard to waste disposal, the following practical recommendations have been formulated 'with the aim of avoiding planning errors which could lead to damaging consequences for the protection and conservation of the region of the lake':

Solid wastes and substances liable to result in pollution, arising from households, craft, commercial and industrial businesses must be collected and disposed of by deposition at the controlled discharge site of the intercommunal syndicate for the discharge of wastes in the Ardennes (SIDA) at Wiltz.

The number of deposits of inert materials (soil, stones, building materials and debris arising from public and private works) must be reduced to one unit per commune.

Decantation sludge, sewage, faecal matter, liquid manure, animal excrement, stable manure, distillery residues, silage juices and similar substances can only be dispersed over ground used for agricultural, forestry or garden cultivation provided they do not exceed the needs of normal manuring.

There is room for improvement in the collection of waste on tourist routes and in recreation areas.

6.1.5

The Law of 27 May 1961[8] establishes a health protection zone around the Esch-sur-Sûre barrage; this zone consists of two parts.

Under Article 3 of this Law the deposition of waste was prohibited in part 1 of the health protection zone; since the entry into force of the Grand-Ducal Regulation of 21 March 1980,[9] this prohibition also applies to part 2.[10]

6.1.6 Household refuse

Since the acceptance in 1973 of a national plan for waste disposal, 101 communes out of a total of 118 have grouped themselves into four intercommunal syndicates for the collection and disposal of household refuse. Indeed, by 1 January 1979, there were only 17 non-affiliated communes disposing of their refuse by uncontrolled tipping, representing 5% of the total refuse produced.[11]

Table I
Affiliation of communes, population per intercommunal syndicate and method of disposal and quantity of refuse

Intercommunal syndicate	No. of com- munes	Population	Method of disposal	Production of household refuse		
				Tonnes p.a.	%	KG/pers. p.a.
SIDA (Wiltz)	16	17,601	controlled tipping	4,541	3.9	258
SIDEC (Diekirch)	26	30,451	controlled tipping	8,160	7.0	268
SIGRE (Flaxweiler)	23	32,560	controlled tipping	9,214	8.0	283
SIDOR (Leudelange)	36	237,813	incineration/ pyrolysis	88,000	76.1	370
Affiliated communes	101	318,425		109,915	95.0	345
Non-affiliated communes	17	21,416		5,889	5.0	275
Total	118	339,841		115,804	100	341

SIDA = Intercommunal syndicate of Ardennes; SIDEC = Intercommunal syndicate of the Diekirch-Ettelbruck-Colmar region; SIGRE = Intercommunal syndicate of the Greven-macher region; SIDOR = Intercommunal syndicate for the destruction of refuse in the cantons of Luxembourg, Esch-Alzette and Capellen.

6.1.7 Industrial waste

Since January 1975 an 'Exchange for Sub-Treatment and Recycling' has existed in the Luxembourg Office of Productivity. This exchange enables the quantity of industrial waste to be limited at source by putting producers of industrial waste in contact with interested parties at international level. In this way the uncontrolled disposal of used oils has been reduced from 34% in 1976 to 12% (from 854 tonnes/year to 301 tonnes/year.[12]

6.2 THE LAW OF 26 JUNE 1980 CONCERNING THE DISPOSAL OF WASTE[13]

6.2.1 Introduction

The main characteristics of the new Luxembourg law on waste lie in the fact that all types of waste are covered by a single general text.[14] Under the meaning of this Law, waste is taken to include

> all residues from manufacturing, processing or use, all substances, materials, products or more generally, all abandoned goods or goods which the owner intends to abandon or is obliged to part with. (Article 1)

However, certain wastes are excluded from the field of application of the Law:

(a) radioactive waste;[15]

(b) waste arising from prospecting, extraction, treatment and storage of mineral resources, as well as mining operations;

(c) animal carcasses, confiscated meats and meat wastes, the destruction of which is controlled by the Law of 6 September 1962 on the destruction and use of animal carcasses, confiscated meats and meat wastes;

(d) waste-water discharged into watercourses or into installations for waste-water;[16]

(e) gaseous effluent emitted into the atmosphere;[17]

(f) explosive substances;

(g) other wastes subject to specific regulations.

6.2.2 Basic principles

The Law contains the three following principles:

(a) Everyone who produces or holds waste in conditions likely to result in harmful effects on the soil, flora and fauna, to damage sites and landscapes, to pollute the air or water, to give rise to noise or odours and, in general, to damage human health or the environment, is required to ensure that such waste is disposed of in accordance with the provisions of the Law, in conditions which will avoid such effects.

(b) Communes are responsible for collecting and disposing of all waste on their territory. They can call upon third parties to carry out this task on their behalf. Communal administrations can, however, exclude the collection and disposal of certain wastes which, by their nature or volume, cannot be disposed of with household refuse.[18] The holder of such wastes must dispose of them himself or engage a third party to undertake the operation on his behalf.

(c) The collection, transportation, importation, exportation, storage, treatment and disposal of wastes are subject to the obligation to obtain prior authorisation from the Minister responsible, i.e. the Minister of the Environment.

6.2.3 Regulatory framework

Articles 5–9 of the Law fix the regulatory framework covering wastes:

(a) Under Article 5, anyone who collects and transports wastes professionally must obtain prior authorisation from the Minister in charge. Authorisation is granted only if the operations can be carried out without danger to human health or damage to the environment. It can be accompanied by conditions, especially with regard to the technical equipment at the disposal of the applicant. New conditions can be imposed if necessary.

(b) Waste must be disposed of by depositing it at controlled sites with sufficient compacting of accumulated wastes in layers, by composting, incineration or any other equivalent process recognised by the competent minister, following prior application from the disposer.

The disposal of waste also involves sorting and treatment operations which are necessary for the recovery of re-usable elements and materials or energy.

Disposal must be carried out under conditions which will not result in harmful effects to the soil, flora, fauna, damage to sites or the countryside, pollution of the air or water, noise and odour which could generally damage human health and the environment (Article 6).

Public and private uncontrolled discharges on refuse tips which, at the time when the Law came into force, were the subject of an authorisation under the legislation on nature conservation and dangerous, dirty and noxious installations, must be closed down within 2 years from the entry into force of the Law, i.e. July 1982 at the latest.

(c) Article 7 states: 'The establishment of an installation serving to store, treat and dispose of wastes as well as all further modifications or extensions to such an installation are subject to prior authorisation to be granted by the competent Minister. Authorisation is granted only if the installation can function without endangering human health or damaging the environment. It can be accompanied by conditions, especially with regard to the technical equipment and storage facilities available to the applicant. The authorisation decree can require the applicant to restore the land after the authorisation has expired. New conditions can be imposed if necessary.

In spite of authorisations granted under the above-mentioned provisions, the Minister responsible for water and forestry administration can refuse authorisation if the installation is likely to damage the beauty of the countryside or if it constitutes a danger for the conservation of the flora and fauna; similarly, the competent authority under the legislation on dangerous, dirty or noxious establishments can refuse authorisation if the installation can result in danger or inconvenience to the public, the neighbourhood or personnel.'

(d) Anyone importing waste into the territory of the Grand-Duchy of Luxembourg must have prior authorisation from the competent Minister (Article 8). The authorisation is granted only to a holder of an authorisation obtained under Article 5 or 7.

(e) Article 9 obliges all those who collect, transport, import and export wastes, or who operate a waste-storage or treatment installation, to keep a precise record, which must be available to the controlling services, showing the type, quantity, destination or disposal process of all wastes handled.

6.2.4 Abandoned vehicles

Vehicles and trailers found in a public place without a registration plate and indication of the name and address of the owner are treated as waste if there is no sign of theft and if after 8 days during which a removal order from the Burgomaster is visibly attached to the car, no one comes forward.

When such a vehicle or trailer constitutes a hindrance or danger to traffic, it may be impounded at an abandoned vehicle storage site until the period of notice mentioned above has expired.

The provisions of Article 7 are applicable to installations used for the storage of abandoned vehicles and tyres.

6.2.5 Agricultural waste

Decantantion sludge, sewage, faecal matter, liquid manure, animal excrement, stable manure, distillery residues, sludge juices and similar substances can only be spread over land used for agricultural, forestry or garden cultivation and provided that they do not exceed the needs of normal manuring. Except when used for this purpose, these substances should be considered as waste.

6.2.6 Special waste

In accordance with the provisions of Article 12, Grand-Ducal Regulations can decree special methods by which certain particularly harmful wastes must be disposed of.

At present there are three Grand-Ducal regulations of this kind, controlling the disposal of used oils, wastes arising from titanium dioxide and polychlorobiphenyls and polychloroterphenyls (see 6.3, 6.4 and 6.5).

In order to facilitate the disposal of particularly harmful wastes, a Grand-Ducal regulation can fix conditions to which the manufacture, importation, warehousing and sale in any form, as well as products generating wastes of this kind, must comply.

Holders of these particularly harmful wastes or of wastes excluded from the communal collection service, see 6.2.2 above, are obliged either to keep a register if they themselves undertake disposal, or to keep collection receipts supplied by the third party they engage to undertake disposal operations.

These registers and receipts provide information on the nature of the waste, the quantity and method of disposal used.

6.2.7 Control and monitoring

Infringements of the Law and implementing regulations are investigated and verified by the officers of the judicial police, agents of the gendarmerie and local police as well as by experts and agents specially appointed for this purpose (Article 14, first paragraph; Grand-Ducal Regulation of 8 May 1981).[19]

In carrying out their legal functions, the experts and agents appointed in this way are acting as officers of the judical police; their competence extends over the entire territory of the Grand-Duchy.

The officers of the judicial police, the agents of the gendarmerie and the police, as well as the specially appointed experts and agents, have access to the premises, lands and means of transport of the people and businesses which are subject to the Law and regulations issued in its implementation.

They can enter such premises, lands and means of transport, but not residential premises, even during the night, if there are serious grounds for suspecting that an infringement of this Law is being committed. They must always inform the head of the establishment, or person in charge, of their presence and he has the right to accompany them during their visit.

They can demand to see the register and documents relating to the wastes. They can also take samples for examination or analysis. When samples are taken a receipt is given and a part of the sample, sealed up, is given to the owner or holder unless he has expressly stated that he renounces this right.

All owners or holders of wastes are required, on demand by the agents, to assist them in their operations.

In the event of a conviction the costs arising from the measures taken under this article are charged to the offender. In all other cases these costs are borne by the State.

6.2.8 Sanctions

Article 18 of the Law distinguishes between two categories of infringement.

Serious infringements are punished by imprisonment of 8 days to 6 months and by a fine of 2,501–200,000 Luxembourg Francs, or by one of these punishments only. Minor infringements are punished by imprisonment of 1 day to 7 days and/or a fine of from 250–2,500 Luxembourg Francs.

In the event of a repeated offence within 2 years of a conviction, the punishments can be increased to double the maximum.

6.3 THE DISPOSAL OF USED OILS

6.3.1 Introduction

The disposal of used oils is controlled by the Grand-Ducal Regulation of 26 June 1980,[20] issued in implementation of Article 12 of the Law of 16 June 1980. This regulation also implements Directive 75/439/EEC of 16 June 1975 of the EEC Council.

The following are considered to be 'used oils': all used semi-liquid or liquid products composed entirely or partially of mineral oil or synthetic oil, including oily tanker residues, oil and water mixtures and emulsions (Article 1).

Disposal of used oils must be carried out by re-use, regeneration, combustion for purposes other than destruction or by any other equivalent process recognised by the Minister of the Environment, following a prior application from the holder.

It is forbidden to add water or any other substance intentionally to used oils before or during collection, to mix oils of different kinds and qualities or to discharge residues resulting from the processing of used oils in an uncontrolled manner.

The holder of used oils must either dispose of them himself, engage an approved collector to undertake this operation, or release them to a third party who is responsible for their disposal.

The communes are responsible for collecting and disposing of used oils within their territory. They can call upon approved collectors to carry out this task on their behalf.

Communes which themselves carry out collection and disposal must collect all the used oils offered in minimum quantities of 200 litres, within 15 days from receipt of a collection request.

An approved collector who carries out the collection and disposal on behalf of the commune must commit himself to respecting the conditions contained in the above paragraph.

6.3.2 Approval as a collector

In order to be approved as a collector of used oils, a person must have sufficient technical means available, either personally or through a contract with a third party, to guarantee the collection, transport, storage and non-polluting use of used oils in accordance with the laws and regulations.

The application for approval must be sent by registered letter to the Minister of the Environment. It must identify the applicant and be accompanied by documents establishing compliance with the requirements of the Grand-Ducal regulation.

The applicant is notified of the decision by registered letter. Reasons are given in the event of refusal. Approval can be given only for an area of at least one commune. The activity of one or more approved collectors in the same territory does not constitute a reason for refusal.

All holders of an approval as a collector can also obtain approval as an importer or exporter on application.

Approval can be suspended or withdrawn if the holder does not fulfil the conditions or if he does not respect the regulatory provisions or special conditions laid down by the Minister.

Physical or legal persons (corporations) who, at the time when the Law came into force, collected, imported or exported used oils can pursue their activities until a decision has been taken regarding the approval application which they must submit within 3 months from the entry into force of the Law.

6.3.3 Registers and statements

The communes, approved collectors, importers and exporters must keep an up-to-date register to be held for 3 years, indicating:

(a)   the daily quantities of used oils collected;

(b)   the use to which these oils have been put, with a precise indication of the quantities sold or exported to different physical and legal persons, including their names or titles and addresses of head offices;

(c)   the quantities which they themselves have regenerated, burnt or processed.

The communes or approved collectors must give a receipt to the person from whom used oil has been collected. This receipt, which must be kept for a period of 3 years, indicates:

(a)   the name or title and address of head office of the collector;

(b)   the name or title and address of head office of the physical or legal person who has given them the used oils;

(c)   the date and place of collection;

(d)   the quantity (in litres) of used oils collected.

A copy of this receipt is held by the commune or approved collector for a period of 3 years and is available to the approved agents and experts.

6.4 THE DISPOSAL OF WASTES ARISING FROM THE TITANIUM DIOXIDE INDUSTRY

A second Grand-Ducal Regulation of 26 June 1980[21] which is also based on Article 12 of the Law of 26 June 1980 and implements EEC Directive 78/176/EEC of 20 February 1978, controls the disposal of wastes arising from the titanium dioxide industry.

Under the meaning of this regulation:

waste is taken to be all residues arising from the titanium dioxide production or treatment process;

disposal is taken to be the collection, sorting, transport and treatment of wastes as well as their storage and deposition on or in the ground and their injection into the ground.

The holder of wastes of this kind must either dispose of them himself or engage a third party to do so on his behalf.

Prior authorisation from the Minister of the Environment is necessary for the discharge, dumping, storage, deposition and injection of wastes.

The authorisation can be granted for a limited period only; it can be renewed. The applicant is notified of the decision by registered letter. Reasons are given in the event of refusal.

The Minister responsible can grant the authorisation only provided that:

(a) the waste cannot be disposed of by more appropriate means;

(b) an assessment carried out on the basis of available scientific and technical knowledge foresees no immediate or future harmful effects on the aquatic environment, the groundwaters, the land or the atmosphere;

(c) no damage results to navigation, fishing, leisure, the extraction of raw materials, desalination, fish breeding, shell-fish breeding or plants, animals or regions of special scientific interest and other legitimate uses of the environment in question.

Whatever the method or degree of treatment of the wastes, their discharge, dumping, storage, deposition and injection must be accompanied by monitoring with regard to the wastes as well as the environment concerned, in all its physical, chemical, biological and ecological aspects.

6.5 THE DISPOSAL OF POLYCHLORINATED BIPHENYLS AND POLYCHLORINATED TERPHENYLS

6.5.1 Introduction

A third Grand-Ducal Regulation of 26 June 1980[22] concerns the disposal of polychlorinated biphenyls, polychlorinated terphenyls and mixtures containing these substances (PCBs and PCTs). It implements EEC Directive 76/403/EEC of 6 April 1976.

Disposal of used PCBs and PCTs, and PCBs and PCTs contained in disused objects or equipment, must be carried out by an establishment or business specially approved for this purpose.

The holder of used PCBs and PCTs must release them to an approved third party for their disposal. Communes are exempt from the collection and disposal of PCBs and PCTs arising from third parties.

6.5.2 Approval to dispose of used PCBs and PCTs

The system is similar to that relating to the approval of collectors of used oils. In order to obtain approval to dispose of used PCBs, and PCTs, a person must himself, or by contract with a third party, have available sufficient technical means to guarantee the non-polluting collection, transport, storage, regeneration, destruction and use of used PCBs and PCTs in accordance with the laws and regulations.

The application for approval must be sent by registered letter to the Minister of the Environment. The applicant is notified of the decision by registered letter. Reasons are given in the event of refusal.

The physical or legal persons who, when the Law came into force, collected, imported or exported used PCBs and PCTs are authorised to pursue their activities until a decision is taken on the approval application which they must submit within 3 months from the entry into force of the Law.

All those approved for the disposal of used PCBs and PCTs can obtain approval as importers or exporters on application.

Approval can be suspended or withdrawn if the holder does not fulfil the conditions or if he does not respect the regulatory provisions or special conditions laid down by the Minister.

6.5.3 Registers and statements

The provisions of the Regulation on PCBs and PCTs on this matter are identical to those on used oils (see 6.3.3 above).

Notes

1. The Law of 27 July 1978 amends the Law of 29 July 1965 concerning the conservation of nature and natural resources; a coordinated text was published in *Mémorial* on 3 October 1978.
2. See Chapter 2.
3. This is mainly covered in the Law of 12 June 1937 concerning the development of towns and other important built-up areas, *Mémorial*, 7 August 1937; the Law of 20 March 1974 concerning land-use planning in general, *Mémorial*, 23 March 1974; and the decision of the Government in Council on 11 November 1977 ruling a policy programme for land-use planning, *Mémorial B*, 27 November 1977.
4. *Mémorial*, 12 June 1961.
5. Decision of the Government in Council on 11 November 1977, *Mémorial B*, 27 November 1977.

6. Article 11, Law of 20 March 1974.
7. *Mémorial B,* 4 August 1978.
8. *Mémorial,* 12 June 1961.
9. *Mémorial,* 15 April 1980.
10. Article 2 (g), Grand-Ducal Regulation, 21 March 1980.
11. Source: Budgetary Report, Minister of the Environment, November 1979, p. 66.
12. Source: *ibid.* p. 69.
13. *Mémorial,* 16 July 1980.
14. See the German Law of 7 June 1972.
15. See Chapter 7.
16. See Chapter 5.
17. See Chapter 3.
18. These wastes will be specified in a decree to be issued by the Minister of the Environment.
19. *Mémorial,* 1981, p. 758.
20. *Mémorial,* 16 July 1980.
21. *Mémorial,* 16 July 1980.
22. *Mémorial,* 16 July 1980.

7
Ionising Radiation

7.1 INTRODUCTION

The outline Law of 25 March 1963 relating to the protection of the population against the dangers resulting from ionising radiation[1] enables the production, manufacture, importation, transport, sale, storage and general use of equipment or substances capable of emitting ionising radiations, as well as the disposal of radioactive substances,[2] to be controlled 'for the purpose of protecting public health'.

As a result of this Law, the Grand-Ducal Decree of 8 February 1967[3] was issued, containing provisions relating to the control of establishments engaged in the importation, distribution and transportation of radioactive substances and the protection and safety of the population as a whole.

7.2 THE CONTROL OF ESTABLISHMENTS

7.2.1 The authorisation procedure

Establishments where equipment or substances capable of emitting ionising radiation are produced, manufactured, stored, sold or used are classified into four categories (see the Annex to this chapter).[4] The heads of establishments in classes I, II and III are required to declare their establishments to the controlling authority and to obtain prior authorisation from that authority. Establishments in class IV are not subject to any authorisation formality under this regulation.[5]

Authorisation is granted by the Government in Council for class I es-

319

tablishments, by a joint decision of the Ministers of Public Health, Labour and Justice for class II establishments, and by the College of Burgomaster and Aldermen of the commune in which the establishment is located for class III establishments.[6]

Establishments including installations which come under different categories are subject to the provisions relating to the highest class involved.

Procedural details are precisely laid down in the regulation.

The authorisation application must contain a great deal of information, including the nature and purpose of the establishment; the individuals in charge of medical and physical control; the obligation to cover civil liability resulting from nuclear activity through an insurance policy or any other financial guarantee; the proposed number of people to be employed in the establishment; and measures for the collection, purification and disposal of eventual radioactive wastes.[7]

There is a public enquiry but it is limited to a duration of only 15 days, which greatly reduces its effectiveness.

Authorisations can be granted for an unlimited duration or for a specified period only. They can also be transferred from one operator to another provided that the transfer is notified immediately to the controlling authority and that the authorisation conditions are observed.

Authorisations relating to class I and class II establishments give the applicant the right to undertake on his own responsibility construction work and to proceed with installations in conformity with the terms of the authorisation granted.

The protection arrangements and installations must be approved by a specially appointed commission or organisation approved by the Minister or Ministers in charge. This approval will depend on conformity with the regulatory provisions and particular conditions imposed on the establishment by the authorisation decision. A written report is prepared and sent to the applicant by the commission or organisation.

The starting up of an installation can take place only if the written report of the commission or organisation is wholly favourable and gives formal authorisation for the starting up operation.

The head of the establishment is required to inform the Minister of Public Health by registered letter of the date fixed for the starting up operation, at least 15 days beforehand.

7.2.2 Emergency measures. Suspension and withdrawal of the authorisation

Article 3 of the Law of 25 March 1963 states that

> when an unforeseen event endangers the health of the population by ionising radiations, the Minister of Public Health will decree measures to be imposed on the producers, manfacturers, transporters, salesmen, holders or users of equipment or substances capable of emitting ionising radiation.

> In the same circumstances the Minister of Public Health will issue the proper measures to avert the dangers which could result from accidental contamination of the area, substances or products by radioactive matter.

> The ministerial decrees issued in implementation of this article become null and void if they are not confirmed within 3 months by a public administration regulation.

In addition to any emergency measures taken by the Minister of Public Health under Article 3, the authority which granted the authorisation can suspend or withdraw that authorisation if, according to written reports from monitoring organisations, the provisions of this regulation and the conditions attached to the authorisation are not observed.

The suspension or withdrawal of an authorisation is communicated to the party concerned, to the authorities consulted at the time of the authorisation decision and to the monitoring organisations.

7.2.3 Alterations to the establishment

All plans for altering or enlarging an establishment which could result in a modification of the nature of the radiation, the protection arrangements or degree of risk, are subject to a decision from the controlling authority according to the authorisation procedure for the category to which the altered or enlarged establishment belongs.

In addition, any cessation of activity must be declared to the authority which granted the authorisation which will then fix the necessary sanitary protection conditions to ensure disposal or re-utilisation of the sources of radiation.

If the head of the establishment or the person in charge of the liquidation of the establishment is unable to satisfy these conditions, the Minister of Public Health can order the seizure of radioactive substances or equipment, and depending on the circumstances, place them under sequestration, without prejudicing the application of sanctions provided for under the regulation.

7.3 THE IMPORTATION, DISTRIBUTION AND TRANSIT OF RADIOACTIVE SUBSTANCES

7.3.1 Introduction

The importation, distribution and transit of radioactive substances, excluding substances as in class IV, can be carried out only by individuals and companies authorised by the Minister of Public Health.[8]

Authorisation is granted for a limited duration and can be general or specific.

Processing of imported, or distributed radioactive substances, or those in transit, must conform in every way to the conditions arising from the regulation and the authorisation. The importer or person responsible for the transit must ensure that the foreign supplier has taken all necessary precautions to make sure that all the conditions are observed.

7.3.2 Importation

The authorisation can impose additional conditions to those contained in the regulation. It can restrict the quantities and activites of the substances imported.

It can specify the substances, quantities and activities for which notification must be given to the Minister of Public Health prior to importation. At any time this authorisation can be withdrawn by justified decision from the Minister of Public Health.

The authorised importation of radioactive substances can only be carried out through customs offices specially appointed for the purpose in the authorisation decree issued by the Minister of Public Health in agreement with the Minister for the Treasury.

An importer who holds a general authorisation must report every month to the Minister of Public Health on the imports he has made.

7.3.3 Distribution

Distributors must be in possession of a declaration from the consignee which states that the necessary authorisations have been obtained.

If the consignee does not satisfy these conditions, the distributor cannot deliver the equipment or substances in question to him.

The distributor must keep detailed records of all deliveries made.

7.3.4 Transit

The authorisation for radioactive substances in transit through the Grand-Duchy of Luxembourg is granted to the applicant by the Minister of Public Health on presentation of a certified copy of:

(1) the export authorisation granted by the responsible authority in the country of origin;

(2) the import authorisation granted by the responsible authority in the country of destination;

(3) the transit authorisation granted by the neighbouring countries through which the package or convoy must pass;

(4) the insurance policy or testimonial that the transporter has a financial guarantee covering his civil liabilities resulting from the proposed operation.

Any transit operation not conforming with these conditions is either refused authorisation by the Minister of Public Health or made subject to certain conditions fixed in the authorisation.

The Minister of Public Health, with the agreement of the Minister of the Treasury, can appoint customs offices through which transit operations can be carried out.

7.4 THE TRANSPORT OF RADIOACTIVE SUBSTANCES

7.4.1

Quite apart from the legal or regulatory provisions controlling transport by land, air or sea, and the international agreements and conventions on the subject,[9] all transporters of radioactive substances—except those involving substances as in class IV—must hold a general or specific authorisation granted jointly by the Minister of Public Health and the Minister of Transport.

The conditions demanded by this authorisation concern the maximum admissible contamination dose; the marking, screening and condition of pacakages; the stowing; the irradiation level at 0.1 m from the surface of the package or transport vehicle, or on its surface; the conditions of transport, security and escort; the protection of employees and travellers; measures to be taken in the event of accident; and conditions of insurance or financial guarantee covering the civil liabilities resulting from the proposed operation and the conditions of the route.

7.4.2

The construction of an engine or vehicle propelled by nuclear energy is subject to joint authorisation by the Minister of Public Health and the Minister of Transport. The procedure to be followed is the same as that for class I establishments.

The movement or parking of an engine or vehicle propelled by nuclear energy on or above Luxembourg territory is subject to prior authorisation from the Minister of Public Health, who can impose special conditions concerning the warehousing, escort, route, docking, landing, parking, security and insurance policy or financial guarantee covering the civil liability involved.

7.5 THE PROTECTION AND SAFETY OF THE POPULATION AS A WHOLE

7.5.1

It is forbidden:

(a) to use equipment giving rise to ionising radiations in the retail footwear trade;

(b) to add radioactive substances to foodstuffs, beauty products, cosmetics, toys or other products for domestic use;

(c) to use photoluminescent sources based on radioactive substances. This prohibition does not apply to luminous sources for activites or in concentrations envisaged for class IV establishments;

(d) to import, hold or transport equipment and products mentioned in (a), (b) or (c).

7.5.2

The treatment of foodstuffs or of medicines with the aid of ionising radiations, and the importation, storage and transportation of these products are subject to special authorisation for each product from the Minister of Public Health. This authorisation can be refused or withdrawn at any time if the conditions laid down by the Minister of Public Health are not respected.

The use of radioactive substances for medical purposes is strictly reserved for individuals holding a doctorate of medicine, or a surgery or maternity qualification, a doctorate of dental medicine or a qualification in veterinary medicine. These doctors must be approved by the Minister of Public Health to be authorised to use radioactive substances for medical purposes.

The monitoring of radioactivity in the territory as a whole is the responsibility of a qualified expert attached to the Medical Director of Public Health.

Under normal conditions, this involves the regular measurement of radioactivity in the air, water, on land and in the food chain, the study of measures to be taken and the coordination of intervention provisions

325

in the event of accidents, as well as the assessment and monitoring of radiation doses received by the population.

7.5.3

As a result of Article 5.2 of the Grand-Ducal Regulation of 8 February 1967, the following safety measures should be applied:

equipment emitting ionising radiations, containers or packages containing radioactive substances, vehicles transporting radioactive substances, deposits of radioactive substances or waste in class I, II or III establishments must carry the symbol and indications described in annex 5 of the regulation;

controlled zones must be identified to the public with the same symbols and indications, and access to these controlled zones is prohibited to the public;

the heads of establishments in class I, II or III must take all necessary measures to prevent the theft or loss of radioactive substances. Any theft or loss of radioactive substances must be declared immediately to the medical inspector of public health for the area who, in conjunction with the police authorities, will take the necessary steps to trace these substances;

the heads of establishments in classes I, II or III must take the necessary steps to prevent accidents and their consequences.

All cases of accidental or deliberate over-exposure must be recorded in the irradiation file and declared immediately to the Inspectorate of Labour and Mines according to a procedure to be fixed by the Minister of Labour who then informs the medical inspector for the area. Any accident which may increase the radiation dose received by the population outside the establishment, either by external or by internal irradiation, must be immediately declared to the medical inspector of the area who, together with a qualified radioprotection expert, will propose the necessary emergency measures to the Minister of Public Health to prevent or restrict further irradiation and damage.

7.6 SANCTIONS

Infringements of the Grand-Ducal Regulation of 8 February 1967 are punishable by imprisonment from 8 days to 1 year and a fine of 500–

50,000 Luxembourg Francs, or by one of these punishments alone, without prejudice to any more severe punishments provided for under other legal provisions.

7.7 INTERNATIONAL CONVENTIONS

The Grand-Duchy is a signatory to the Convention on Civil Liability in the field of nuclear energy (Paris, 29 July 1960)[10] and to the Convention relating to the liability of the operator of a nuclear installation (Brussels, 31 January 1963).

ANNEX
Classes of Establishment

Class I

(1) Establishments including one or more nuclear reactors.

(2) Establishments involving the presence of irradiated nuclear fuel.

(3) Establishments involving the presence of fissile substances in such conditions that half the minimum critical mass could be exceeded.

Class II

(1) Establishments involving the presence of radioactive nucleides, the total activity of which is given by the value X_2 of Table B, Annex 2 of the regulation, excluding states, quantities and activities included in Class I.

(2) Establishments engaged in the collection, treatment, processing or storage of radioactive waste.

(3) Establishments involving the habitual presence of equipment producing X-rays which can function under a point voltage of more than 200 kV.

(4) Establishments involving the presence of particle accelerators.

(5) Radioactive equipment and products used in a mobile manner, if they are established permanently, even if they involve the presence of nucleides included in class III.

Class III

(1) Establishments involving the presence of quantities of radioactive nucleides, the total activity of which is given by the value X_3 of Table B, Annex 2 of the regulation, excluding states, quantities and activities included in class I or II.

(2) Establishments involving the habitual presence of equipment producing X-rays which can function under a tension equal to or less than 200 kV.

Class IV

(1) Establishments involving the presence of radioactive nucleides, the total activity of which is given by the value X_4 of Table B, Annex 2 of the regulation, excluding states, quantities and activities included in class I, II or III.

(2) Establishments involving the habitual presence of equipment which, without being X-ray producing, emits X-rays (e.g. television apparatus), excluding equipment in class II.

(3) Establishments involving the presence of equipment containing radioactive substances, the total activity of which exceeds the values fixed under category (1) of this class, on condition that:
   (a) it is sealed;
   (b) the likely dose does not exceed 0.1 mrem/hour at any point more than 0.1 m away from the surface of the apparatus.

(4) Establishments involving radioactive substances in any quantity provided that the concentration of these substances is less than 0.002 µCi/g, or less than 0.01 µCi/g if natural solid radioactive substances are involved.

Notes

1. *Mémorial*, 10 April 1963.
2. *Ionising radiation* is taken to mean electromagnetic radiation (photons or quanta of gamma or X-rays) or corpuscular radiation (alpha or beta particles, electrons, positrons, protons, neutrons and heavy particles) capable of determining the formation of ions. *Radioactive substances* are taken to be all substances which present the phenomenon of radioactivity.
3. *Mémorial*, 8 March 1967.
4. See Annex to this Chapter.
5. There is an exception: the type of equipment under class IV, category (2), provided that they function at a tension point equal to or more than 5 kV and they emit X-rays of such an intensity that the likely dose receivable on their surface exceeds 0.5 mrem/hour, as well as the type of equipment in class IV, category (3), which must

be declared by the manufacturer or importer to the Minister of Public Health and approved by him.

6. At present, nearly 200 authorisations have been granted, of which half are class II establishments and the rest are class III establishments. There are as yet no class I establishments.

7. In practice, radioactive wastes arising from used radioactive substances are returned to the supplier who, up to now, has been a producer in another country.

8. See the recommendation of 18 April 1966 of the Benelux Committee of Ministers: the authorisation for the importation, transit and transportation of radioactive substances or equipment granted by the Belgian or Dutch authorities which also concern Luxembourg territory are recognised to be valid.

9. This refers to the ARD convention (transport by road), RID convention (transport by rail) and IATA convention (transport by air).

10. As well as the additional Protocol (Paris, 28 January 1964).

8

Product Control

8.1 FOODS

8.1.1 General

8.1.1.1

The Law of 25 September 1953 'to reorganise the control of foodstuffs, drinks and commodities'[1] enables the Government to control, monitor and even prohibit, in the interests of public health:

(a) the manufacture, preparation, processing, sale and distribution of:

> foodstuffs and drinks or medicines intended for use by humans or animals;
>
> consumer goods and clothing;
>
> cosmetic products and toiletries;
>
> normal products and articles used in the home such as toys, carpets, furniture, hangings, utensils, paints, essences and other liquid or solid substances;

(b) the sale and distribution of equipment, utensils, containers and other articles used in the manufacture or intended to be placed in contact with foodstuffs, drinks or medicines, consumer goods, cosmetic products and toiletries.

8.1.1.2 MONITORING THE APPLICATION OF THE LAW

Monitoring the manufacture, preparation, processing, sale and distribution of the products listed in 8.1.1 is carried out on the authority of the Minister of Public Health or his delegate, by experts of the respective State control services—the State laboratories—and the agents and officials of the national or local police. They have the right to:

enter premises in which the articles listed under 8.1.1 are manufactured, prepared, processed, stored, displayed, sold or distributed;

to visit and inspect the vehicles and other means of transport which contain or might contain these articles;

to require the production of certain commercial records and documents;

to take samples for examination or analysis;

to seize and, if necessary, sequestrate the listed articles, in addition to articles or materials used or intended for use in their manufacture or sale, which are recognised to be falsified, adulterated or spoiled, as well as commercial records and other documents required under the law and public administration regulations relating to them;

to seize and put out of service articles which constitute a danger to public health because of their unhealthiness as verified by the medical public health inspector (Article 7).

8.1.1.3 SANCTIONS

Articles 9–22 of the Law of 25 September 1953 contain the penal provisions on the subject. The severest punishment is 15–20 years hard labour, if the punishable act has caused the death of an individual.

The confiscation can be ordered of goods which have been the subject of an infringement, or which have served to commit an infringement, or which have been intended to be used to commit an infringement. If the confiscated articles can be used for purposes which are not dangerous or unhealthy, the State Procurator can place them at the disposal of the commune in which the infringement was committed, to be given to hostels, charity offices or other establishments; where this is not the case, the articles must be destroyed.

8.1.2 Preservatives

8.1.2.1

The Grand-Ducal Regulation of 8 June 1977[2] establishes a positive list of preservatives which can be used in foodstuffs intended for human consumption.

It is forbidden to use substances other than those listed in Annex A, parts II and III of the regulation for the preservation of foodstuffs.

The addition of these preservatives to foodstuffs is only authorised for foods listed in Annex C of the Regulation, under conditions fixed for each heading of the Annex.

The substances must conform to the general and specific purity criteria fixed in Annex B of the regulation.

Annexes A, B and C can be altered by ministerial regulation following directives of the Council of Ministers of the European Communities or decisions of the Committee of Ministers of the Benelux Economic Union. All other modifications to these Annexes must be subject of a Grand-Ducal Regulation.

However, the authorisation to use one of the preservatives listed in Annex A of the regulation can be suspended by ministerial regulation in the event of danger to health. Equally, the Minister of Public Health can can authorise the use, for transitional period, of one or several of the substances listed in Annex A in certain foods and drinks which are not included in Annex C, provided that they are not the subject of a special regulation.

8.1.2.2

The smoking of certain foods is only authorised if the smoke is produced by wood or woody vegetation in its natural state, excluding wood and vegetation which has been impregnated, coloured, glued or treated in a similar way, and on condition that no risk to human health arises from the smoke.

8.1.2.3

The regulation specifies in detail the indications to be included on the packaging and container before the product can be sold:

The name and address of the manufacturer established within the European Community. Anyone who imports a product from a third country is considered to be the manufacturer.

The number and name of the substances as listed in Annex A. The name must be given in French and German.

The mention 'for foods (limited use . . .').

In the event of a mixture of substances listed in Annex A:

the name of each of the components, or, where appropriate, their number as given in Annex A;

their percentage, so far as one or more of the substances listed in Annex A is concerned, or if this obligation is provided for under provisions relating to other categories of additives.

8.1.2.4

The manufacture, importation into Luxembourg, exportation to a member state of the European Communities, storage for sale, display and sale of foods and drinks which do not comply with the provisions of the regulation are prohibited.

However, the regulation does not apply to products used for coating foodstuffs, to products used to combat organisms which are harmful to plants and vegetable products, to antibacterial products used for the treatment of water or to anti-oxidant products; nor does it apply to foods exported outside the European Community.

In addition, the use of food products which have preserving properties such as vinegar, sodium chlorite, ethyl alcohol, oils and sugars is not affected by this regulation.

8.1.3 Emulsifying, stabilising, thickening and gelling agents

8.1.3.1

The Grand-Ducal Regulation of 9 October 1979[3] authorises the treatment of foodstuffs for human consumption with emulsifying, stabilising, thickening and gelling agents, provided they are listed in Annex I of the regulation. The addition of these substances is permitted only in foodstuffs listed in Annex III of the regulation and under the conditions listed under each heading of that annex.[4]

The substances listed must comply with general and specific purity criteria.

The general criteria are as follows:

The substances must not contain any dangerous elements from the toxicological viewpoint, particularly heavy metals;

They must not contain more than 3 mg/kg of arsenic or more than 10 mg/kg of lead;

They must not contain more than 50 mg/kg of copper and zinc taken together, and the zinc content must not exceed 25 mg/kg unless an exemption has been granted.

The specific criteria are fixed in Annex II of the Regulation.

The substances listed can be placed on the market only if their packaging or containers carry the indications specified in Article 5 of the Regulation.

8.1.3.2

Similar to the Grand-Ducal Regulation of 8 June 1977 relating to preservatives, this Regulation allows temporary exemptions; the Minister responsible is the Minister of Health.

Annexes I, II and III can be modified by ministerial regulation following directives from the Council of Ministers of the European Communities and decisions of the Committee of Ministers of the Benelux Economic Union. All other modifications to these Annexes must be the subject of a Grand-Ducal regulation.

8.1.3.3

It so forbidden to manufacture, export, store or transport with a view to sale, display, sell, offer as a free or conditional gift or exchange the substances in question if they do not conform to the provisions of the Regulation. The same prohibitions apply for foodstuffs containing substances which do not conform. However, it should bs stated that the prohibition of substances not listed in Annex I of the regulation for the treatment of foodstuffs intended for human consumption does not apply to:

(1) foodstuffs with emlusifying, stabilising, thickening or gelling properties such as eggs, flour, starch;

(2) emulsifiers used in separating products;

(3) acids or bases, and salts which modify or stabilise the pH;

(4) blood plasma, modified starches, edible gelatine and edible soluble proteins and their salts;

(5) products containing pectin and obtained from dried apple peel or

dried citrus skins or a mixture of the two, by diluted acid treatment followed by partial neutralisation with sodium or potassium salts.

8.1.4 Meats

The Grand-Ducal Regulation of 25 February 1980[5] concerning the control of meats and certain foods makes the following subject to control by the services responsible, in the interests of public health:

the transport and ante-mortem inspection of animals to be slaughtered;

the slaughter, inspection and butchery of fresh meats, poultry and game;

the manufacture of meat-based products;

the importation and exportation of fresh meats and meat-based products;

the transport and storage of fresh meats and meat-based products;

the sale and distribution of fresh meats, and meat-based products, poultry and rabbits, game and marine and freshwater products for human consumption;

the hygiene of personnel and hygiene conditions of premises and installations.

8.1.5 Special food

Directive 94/77/EEC of 21 December 1976 has been implemented in internal law by the Grand-Ducal Regulation of 7 November 1980 relating to foodstuffs intended for particular nutritional uses.[6]

8.1.6 Materials and objects in contact with foodstuffs

The Grand-Ducal Regulation of 13 April 1978[7] implements in internal law Directive 893/76/EEC of 23 November 1976 relating to the harmonisation of legislation in member states concerning materials and objects intended to come into contact with foodstuffs. It applies to materials and objects, which, as finished products, are intended to come into contact, or are placed in contact, during use, with foodstuffs,[8] as well as materials and objects in contact with water intended for human consumption.[9]

These materials and objects must be manufactured in accordance with good manufacturing practices so that, in normal or foreseeable conditions of use, they do not transmit constituents to the foodstuffs in a quantity liable to;

> present a danger to human health;
>
> result in an unacceptable alteration in the composition of the food-stuffs or in their organoleptic characteristics.

They must also, when sold, be accompanied by the indications required by Article 3 of the regulation.[10]

8.1.7 Additives in animal foods

As a result of the Grand-Ducal Regulation of 26 November 1979 concerning the use and control of additives in animal foods,[11] it is only permissible to manufacture, prepare, import, export, hold or transport with a view to sale, offer for sale, conditional or free gift or exchange animal foods provided they contain only the additives listed in Annexes I and II of the regulation, and only under the conditions indicated therein.[12]

8.2 NON-FOOD PRODUCTS

8.2.1 Cosmetic products

8.2.1.1

As in Belgian legislation, the Grand-Ducal Regulation of 24 October 1978 relating to cosmetic products uses the method of negative lists.[13]

Without impairing the general requirement of not being likely to harm human health when applied under normal conditions of use, the sale of products containing the following is prohibited:[14]

(a)   the substances listed in Annex II of the regulation;

(b)   the substances listed in the first part of Annex III of the regulation in excess of the limits and outside the conditions indicated;

(c)   dyes other than those listed in the second part of Annex III; if the cosmetic products are to be applied near the eyes, on the lips, inside the mouth or on the external genital organs, these dyes cannot be used in excess of the limits and outside the conditions indicated in the same annex.

In addition, cosmetic products can only be put on the market if their exterior packaging, containers or labels carrying the information required under Article 7 of the regulation 'in indelible and easily visible and legible writing'.

8.2.1.2

When, following the use of a cosmetic product, the need for rapid suitable medical treatment arises, adequate and sufficient information must, on request, be placed at the disposal of the relevant Ministry of Public Health service by the manufacturer and the importer. The service ensures that the information will be used only for the purposes of the treatment.

8.2.1.3

While the marketing of cosmetic products containing oestrogen, oestradiol and its esters, progesterone and ethisterone (hormonal substances) is prohibited in principle, the Minister of Public Health can, on request from the manufacturer or importer, authorise their use in a specific product, provided that the applicant presents a data file to demonstrate that the product is harmless.

On the other hand, the importation manufacture, storage with a view to sale, offer for sale and sale of all products to be used orally, intended exclusively or primarily to colour the skin are absolutely prohibited.[15]

8.2.2 Detergents

The Law of 15 April 1980 approves the European Agreement on the limitation of the use of certain detergents in washing and cleaning products, signed in Strasbourg on 16 September 1968.[16]

8.2.3 Pesticides and agricultural pharmaceuticals

8.2.3.1

The Law of 28 February 1968 to control pesticides and agricultural pharmaceuticals[17] authorises the Grand-Duke, after seeking the opinion of the Medical Council, to control the storage, manufacture, import, delivery or transport with a view to sale, the offer for sale or sale, the free or conditional gift, exchange and use of pesticides and agricultural pharmaceuticals.[18]

The Law prohibits the storage, manufacture, import, delivery or transport with a view to sale, offer for sale or sale, free or conditional gift or exchange of pesticides or agricultural pharmaceuticals not approved by the Government.

A product is approved only if it can be stated with reasonable certainty that it is of a quality to achieve the objective for which it is intended and that its appropriate use will not give rise to any harmful secondary effect.

8.2.3.2 MONITORING THE APPLICATION OF THE LAW

Monitoring the storage, manufacture, import, delivery or transport with a view to sale, the offer for sale, sale, delivery and free or conditional gift, the exchange and use of pesticides and agricultural pharmaceuticals is carried out on the authority of the Minister of Public Health or his delegate with regard to pesticides, and of the Minister of Agriculture or his delegate with regard to agricultural pharmaceuticals.

Article 31 of the Grand-Ducal Regulation of 29 May 1970 (see 8.2.3.3) appoints experts and agents to monitor the provisions of the Law and its implementing regulations. These experts and agents, as well as officials and agents of the general and local police, can:

(a) enter any premises in which the products in question are manufactured, prepared, processed, stored, displayed, sold and distributed at any time that they are open to the public, and even during the night if there are serious reasons to suppose an infringement of this Law and its implementing regulations is being committed;

(b) monitor the use of pesticides and agricultural pharmaceuticals with the exception of the provisions concerning the inviolability of the home;

(c) request the production of all commercial records and documents required by the Grand-Ducal regulations issued in implementation of the Law;

(d) remove at will samples of pesticides and agricultural pharmaceuticals, as well as materials used during their manufacture, for examination or analysis;

(e) seize, and if necessary sequestrate, the products concerned as well as the commercial records and documents required under this Law and the Grand-Ducal regulations issued in its implementation.

8.2.3.3 SANCTIONS

In principle, any infringement of an implementing regulation of the Law is punishable by imprisonment from 8 days to 3 months and a fine of 1,000–50,000 Luxembourg Francs, or by one of these punishments alone.

In addition, the following can be ordered:

the confiscation of the products concerned which have been the subject of the infringement;

the confiscation of illegal profits;

the closure, for a period not exceeding 3 years, of the establishment where the infringement was committed;

the publication of the decrees and judgements in one or more daily newspapers in the Grand-Duchy at the expense of the person who committed the infringement.

8.2.3.4

As a result of Article 3 of the Grand-Ducal Regulation of 29 May 1970 concerning the control of pesticides for agricultural use and agricultural pharmaceuticals,[19] approval is granted by the Minister of Agriculture and the Minister of Public Health, on the opinion of an 'approval commission'. This commission is made up of 6 members appointed by these Ministers, including 3 officials from the Ministry of Agriculture, one of whom takes on the presidency, 2 officials from the Ministry of Public Health and 1 official from the Ministry of Labour and Social Security. The latter is nominated on the proposal of the Minister responsible.

The following, among others, are considered to be harmful secondary effects:

(a) harmful effect on public health;

(b) harm from the point of view of health or any danger threatening the safety of the product user while observing the necessary precautionary measures;

(c) harmful effect on the quality of foods;

(d) harmful effect on the fertility of the soil, plants or parts of plants, as well as of animals whose conservation is desirable, if the damage is disproportionate in comparison with the purpose for which the product is used.

In granting approval, the Ministers of Agriculture and Public Health determine, when necessary, special conditions to which the marketing of the product is subject.

Approval is granted for a maximum duration of 10 years. The Ministers of Agriculture and Public Health can at any time, provided they give at least 6 months' notice, suspend or withdraw the approval without any compensation being payable by the State. No notice is required if the suspension or withdrawal of approval is required for public health reasons.

Reasons for the suspension or withdrawal decision are given to the party concerned.

8.2.3.5

Agricultural pharmaceuticals are divided into four groups according to the degree of danger they present:[20]

List A; particularly dangerous toxic products;

List B: toxic products;

List C: products containing the substances in List A or B in low quantities and concentrations which are not dangerous;

List D: non-toxic products.

The classification of products is carried out by the Minister of Public Health, on the opinion of a commission made up of 6 members nominated by him as follows: 3 officials from the Ministry of Public Health, one of whom takes on the presidency, 2 officials from the Ministry of Agriculture and 1 official from the Ministry of Labour and Social Security. The last three are appointed on the proposal of the Ministers responsible.

This commission can also define the meaning of the terms 'harm' and 'harmful effect'.

8.2.3.6

With regard to the products on Lists A and B, the importation, purchase, storage with a view to sale, offer for sale, sale and free offer are reserved exclusively to pharmacists with laboratories open to the public, and to individuals who are specially approved for the purpose by the Minister of Public Health. These two categories of individuals are known as 'approved sales persons'.

The toxic agricultural chemicals on List A cannot be purchased for use or used except by individuals who are specially approved for the purpose by the Ministers of Public Health and Agriculture. These individuals are known as 'approved users'. These approved users are responsible from the health viewpoint for the use of the products concerned, both with regard to the methods used and the treatment itself.

Chapter VI of the Grand-Ducal Regulation of 29 May 1970 contains a series of provisions relating to conservation and safety measures to be adopted by these approved sales persons and users.

8.2.3.7

Finally, it should be pointed out that the marketing and use of agricultural chemical products based on aldrin (HHDN), dieldrin (HEOD), endrin, heptachlor, chlordane, DDT, hexachlorobenzene and alkyl-mercury compunds are prohibited.[21]

8.2.4 Fertilizers

8.2.4.1

As a result of Article 1 of the Law of 26 February 1973,[22] Grand-Ducal regulations can fix the conditions of composition, quality, packaging, identification, marketing, transport and storage of fertilizers.

Monitoring of the measures implemented by these Grand-Ducal regulations is carried out on the authority of the members of the government responsible for agriculture and economic affairs.

Infringements of the Law and its implementing regulations are investigated by the officers of the civil police and agents of the gendarmerie and police, customs agents, as well as agents appointed under Article 18 of the Grand-Ducal Regulation of 24 January 1979 (see 8.2.4.2):

(a) as experts: engineers of the control and testing laboratories division of the agricultural technical services administration;

(b) as agents: chemists and career agents with a technical diploma from the control and testing laboratories division of the agricultural technical services administration.

For the purpose of investigating and verifying infringements, all stages of the manufacture, marketing and transport are controlled. When there are serious reasons to suppose that an infringement is being committed, the agents can enter the premises, even during the night, where the objects covered by Article 1 are manufactured, prepared, stored, displayed, sold or distributed. However, if a private dwelling is involved, a search warrant is required. The agents can also:

(a) take samples each time they consider it to be necessary. The owner or holder is compensated for the value of the samples at the current price;

(b) require the production of all commercial records relating to the objects, and all documents required by the Grand-Ducal regulations issued in implementation of the Law;

(c) seize and, if necessary, sequestrate the objects concerned, as well as the commercial records and documents required under the Grand-Ducal regulations issued in implementation of the Law.

Infringements of the Grand-Ducal regulations issued under the Law are punishable by imprisonment from 8 days to 5 years and a fine of 500 to 1 million Luxembourg Francs, or by one of these punishments only.

In addition, the confiscation of the products which were the subject of the infringement, as well as the confiscation of illegal profits can be ordered.

8.2.4.2

The Grand-Ducal Regulation of 24 January 1979 relating to the marketing of fertilizers[23] prohibits the marketing of manure, fertilizers and all products which act specifically to stimulate crop production (with the exception of agricultural health products used primarily in the preparation of another product) if these products are not included on the 'positive lists' of annexes I–VII of the regulation.[24]

The products covered by these annexes can be marketed only under the denominations in column (a) of the annexes. They must also conform to the provisions in column (b), to the criteria and other requirements in column (c) and must possess the substantial qualities listed under column (d), the contents of which must be guaranteed.[25]

By means of a waiver of these provisions, the Ministers of Agriculture and National Economy can:

with the exception of EEC fertilizers, under conditions determined by them, permit the marketing of the products included in the annexes of this regulation but which for an accidental reason do not satisfy the provisions contained in the regulation;

under conditions determined by them, permit the marketing of products which are not included in the annexes of this regulation, but the use of which is in the interests of agriculture, bearing in mind their recognised qualities.

But in all these cases the products covered by the regulation must be:

(1)   of unadulterated, marketable and commercial quality and must not have undergone any treatment to alter their nature or quality to the extent that their composition no longer conforms to the normal product;

(2)   in a state ready for use when sold;

(3)   have a degree of homogeneity within limits compatible with the factory manufacturing conditions;

(4)   free from toxic or harmful substances, harmful insects, nematodes, viable rust spores, rot or other phyto-pathological organisms in sufficient quantities to have an undesirable effect on crop cultivation or on the health of humans and animals when these products are used in normal doses and in a judicious manner.

8.2.4.3

Articles 7–17 of the Grand-Ducal Regulation of 24 January 1979 control the problems of identification, of guarantee and of packaging for manure and fertilizers.

The Grand-Ducal Regulation of 29 February 1980 makes applicable in the Grand-Duchy of Luxembourg methods of sampling and analysis for the official control of fertilizers which are contained in EEC Directive 535/77/EEC of 22 June 1977, amended by Directive 138/79/EEC of 14 December 1978, and in decisions M (78) 10 of 14 November 1978 and

M (79) 2 of 4 May 1979 of the Committee of Ministers of the Benelux Economic Union.

8.2.5 Medicinal and toxic substances

The Law of 19 February 1973 on medicinal substances and the fight against drug addiction enables the Grand-Duke, in consultation with the Medical College, to control:[26]

(a) the manufacture, wholesale storage and sale of medicinal substances;

(b) the import, export, manufacture, transport, storage, sale and offer for sale, the delivery or purchase, as a free or conditional gift and the use of narcotics, bacterial cultures and toxins, toxic, soporific, pyschotropic, disinfectant and antiseptic substances, as well as the cultivation of plants from which these substances can be extracted;

(c) the inspection and revision of pharmacies and medicinal stores, and of the businesses under (a) and (b), as well as the taking of samples, the seizure and destruction of substances which have been altered or illegally stored.

8.2.6 Dangerous substances

The Law of 14 March 1979[27] fixes the rules relating to the classification, packaging and labelling of dangerous substances marketed in Luxembourg.[28]

Annex I of the Law reproduces the list of dangerous substances classed according to the greatest degree of danger and the specific nature of the risks.[29]

Dangerous substances can be marketed only if they conform to the provisions of the Law of 14 March 1979.

As regards solidity and water-tightness, their packaging must comply with the following conditions:

(1) the packaging must be organised and closed in such a way as to prevent any loss of content, with the exception of regulatory safety devices;

(2) the material of which the packaging and closure are made must not

be damaged by the contents, nor liable to form harmful or dangerous combinations with the latter;

(3) the packaging and closure must, in all parts, be solid and strong
enough to exclude any possibility of release, and to be guaranteed
to fulfil normal handling requirements.

In addition, the packaging must carry the following information 'in a
legible and indelible manner':

the name of the substance;

the origin of the substance;

the danger symbols and information relating to the use of the
substance;

a reminder of the particular risks arising from these dangers.

Monitoring of the application of the provisions of the Law and its
implementing regulations is carried out by the Inspectorate of Labour
and Mines.

Infringements are punishable by imprisonment from 8 days to 1 month,
and a fine of 2,500–250,000 Luxembourg Francs, or one of these punishments only.

In addition, the confiscation of the substances which were used to commit the infringement can be ordered by the courts.

Notes

1. *Mémorial*, 10 October 1953, p. 1259; see also the Law of 9 August 1971 completing
   the Law of 25 September 1953 to reorganise the control of foodstuffs, drinks and
   normal products, *Mémorial*, 24 September 1971, p. 1781.
2. *Mémorial*, 6 July, 1977, p. 1020, This Regulation was issued in implementation of
   the Law of 25 September 1953 and a series of EEC Directives. Article 6 was modified
   by the Grand-Ducal Regulation of 26 May 1979, *Mémorial*, 28 June 1979, p. 1049.
3. *Mémorial* 1979, p. 1778. This Regulation was issued in implementation of the Law
   of 25 September 1953 and Directives 329/74/EEC, 612/78/EEC and 663/78/EEC of
   the Council of the European Communities. Annexes I and III of this Grand-Ducal
   Regulation have been amended by the Grand-Ducal Regulation of 8 October 1980,
   Mémorial, 20 October 1980.
4. Emulsifying agents and stabilising agents are taken to mean substances which, when
   added to a food, enable the uniform dispersion of two or more non-mixing phases
   to be achieved or maintained. Thickening agents are substances which, when added
   to a food, increase its viscosity. Gelling agents are substances which, when added to
   a food, give it the consistency of a jelly.
5. *Mémorial*, 6 May 1980, p. 670.
6. *Mémorial*, 1980, p. 1986.
7. *Mémorial*, 5 May 1978, p. 476. See also the Ministerial Regulation of 24 October

1980 which determines the symbol to accompany the materials and objects intended to come into contact with foodstuffs, *Mémorial*, 22 November 1980, p. 1981.

8. Coating or glazing materials—such as materials for coating cheeses, pork products or fruits—which are in contact with foodstuffs and are liable to be consumed with the foods are not subject to the provisions of the Regulation of 13 April 1978.

9. The Regulation does not, however, apply to public or private fixed installations used for the distribution of water.

10. See also the Grand-Ducal Regulation of 19 November 1974 concerning the labelling and packaging of foodstuffs, *Mémorial*, 4 December 1974, p. 1734.

11. *Mémorial*, 1979, p. 2294. This Regulation was issued in implementation of the Law of 25 September 1953 and of Directive 524/70/EEC of 23 November 1970, as subsequently modified.

12. Annex I contains the permitted additives for exchanges between the member countries of the European Communities; Annex II lists the additives which are permitted in excess of the national level.

13. *Mémorial*, 4 December 1978, p. 1936. This Regulation was issued in implementation of the Law of 25 September 1953 and Directive 768/76/EEC of 27 July 1976. It was modified by the Grand-Ducal Regulation of 5 March 1979, *Mémorial*, 21 March 1979, p. 411. See also the Ministerial Regulation of 3 December 1979, *Mémorial*, 22 December 1979, p. 1875. The definition of 'cosmetic product' is identical to that contained in the Belgian Royal Decree of 10 May 1978. Annex I contains an 'indicative list by category' of cosmetic products:

creams, emulsions, lotions, jellies and oils for the skin (hands, face, feet, etc.)

face masks (excluding products for the superficial abrasion of the skin by chemical means)

coloured foundation products (liquid, cream, and powder)

face powder, talcum powder, etc.

toilet soap, deodorising soap, etc.

perfumes, toilet waters and eau de cologne

bath and shower preparations (salts, foam, oil, etc.)

depilatory products

deodorants and antiperspirants

hair care products:

    hair tints and dyes

    perming, waving and straightening products

    setting products

    cleaning products (lotions, powders, shampoos)

    products for maintaining the hair (lotion, lacquer, brilliantine)

shaving products

makeup and makeup removing products for the face and eyes

products to be applied to the lips

products for the care of the mouth and teeth

nail care products

sun products

products for tanning without sun

products for bleaching the skin

anti-wrinkle products

14. Except for temporary exemptions provided for under Articles 4 and 5 of the Regulation.
15. Grand-Ducal Regulation of 18 April 1978 prohibiting the marketing of products for oral use, exclusively or partially intended to alter the colour of the skin, *Mémorial*, 8 May 1978, p. 491.
16. *Mémorial*, 30 April 1980, p. 481. All member states of the EEC, except Ireland and Greece, are contracted parties to this agreement.
17. *Mémorial*, 12 March 1968, p. 123.
18. Pesticides are taken to be substances and preparations for the purpose of destroying or preventing the action of harmful animals, plants, micro-organisms or viruses. They are subdivided into pesticides for agricultural use and pesticides for non-agricultural use.

    Agricultural chemicals are taken to be:

    (a) pesticides for agricultural use;
    (b) substances and preparations to stimulate or regularise crop production or to ensure the conservation of crops or parts of crops;

    (c) substances and preparations for the destruction of weeds;

    (d) substances and preparations for the destruction of leaves or the prevention of undesirable growth;

    (e) Micro-organisms and viruses as active agents in the fight against parasites;

    (f) moisturisers and adhesives intended to encourage the action of the substances and preparations under points (a)–(d), provided they are marketed for this purpose.

19. This Regulation is not applicable:

    (1) to the manufacture, preparation, storage, wholesale trade and retail trade of medicinal substances and pharmacuetical specialities. See 8.2.5;

    (2) to manure and fertilisers. See 8.2.4.1;

    (3) hydrocyanic acid and substances likely to produce or release hydrocyanic acid, provided the provisions concerning the control of certain establishments reputed to be dangerous, dirty or inconvenient apply to them. See Chapter 2;

    (4) to agricultural chemicals in transit or intended for export to countries other than Benelux, provided they are accompanied by documentary proof, or, if they are located in factories, workshops, shops, depots and warehouses, that each package carries the word 'export' clearly visible, that they are placed in an area exlusively used for this purpose and that the owner or holder can provide documentary proof of this use;

    (5) to agricultural chemicals intended for experimental or scientific use, provided

that the individual using the products for this purpose has obtained authorisation from the Minister of Agriculture.

20. See the Ministerial Regulation of 4 July 1974, *Mémorial*, 5 August 1974, p. 1328.

21. Grand-Ducal Regulation of 29 March 1975, *Mémorial*, 29 March 1975, p. 422; see also *Mémorial*, 26 July 1977, p. 1288.

22. Law of 26 February 1973 controlling the manufacture and marketing of manure, *Mémorial*, 15 March 1973, p. 382.

23. *Mémorial*, 1979, p. 244.

24. These annexes can be modified and extended by ministerial regulation.

25. Only products contained in Annexes I and II can be marketed as 'EEC fertilizers', and only if they comply with the criteria and other requirements fixed for such fertilisers in this Regulation and its Annexes I and II.

26. *Mémorial*, 28 March 1980, p. 318; see also the Grand-Ducal Regulations of 4 March 1974 and 28 November 1980 concerning certain toxic substances, *Mémorial*, 11 December 1980, p. 2060.

27. *Mémorial*, 23 March 1979, p. 424. The provisions of this Law do not apply to:

(a) medicines, narcotics and radioactive substances;

(b) the transport of dangerous substances by rail, road, river, sea or air;

(c) munitions and objects containing explosive materials such as fuels and inflammatory substances.

28. Under the meaning of the Law, the following definitions are used:

(1) *substances:* chemical elements and their compounds as they are present in the natural state or as they are produced by industry;

(2) *preparations:* mixtures or solutions made up of two or more substances. Under the meaning of the Law, substances and preparations are dangerous, as follows:

(i) *explosives:* substances and preparations which could explode under the effect of flame or which are more sensitive to shock or jolting than dinitrobenzene;

(ii) *combustible substances:* substances and preparations which, in contact with other substances, particularly inflammable substances, present a very exothermic reaction;

(iii) *easily inflammable:* substances and preparations which can become hot and finally burst into flame at normal temperature without additional energy, or

solids which can easily burst into flame by the brief action of an inflammatory source and which continue to burn or be consumed after the source of flame has been removed, or

liquids of which the flash point is lower than 21°C, or

gases which are inflammable with humid air, developing easily inflammable gas in quantities which are dangerous;

(iv) *inflammable:* liquid substances and preparations with a flash point between 21°C and 55°C;

(v) *toxic:* substances and preparations which, when inhaled, swallowed or absorbed through the skin can lead to serious, acute or chronic risks, or even death;

(vi)   *harmful:* substances and preparations which, through inhalation, swallowing or absorption through the skin can lead to limited serious risks;

(vii)   *corrosive:* substances and preparations which, in contact with living tissue, can have a destructive action on the latter;

(viii)   *irritant:* non-corrosive substances and preparations which, by immediate, prolonged or repeated contact with the skin or mucous membranes, can cause inflammation.

29. Annex I: list of dangerous substances classed according to atomic number of the most characteristic element of their property (EEC *OJ* L360, 30 December 1976);

Annex II: danger symbols (EEC *OJ* L167, 25 June 1973);

Annex III: nature of particular risks of dangerous substances (EEC *OJ* L360, 30 December 1976);

Annex IV: recommendations for caution concerning dangerous substances (EEC *OJ* L360, 30 December 1976);

Annex V: equipment and methods of determining the flash points of liquid substances and preparations (EEC *OJ* L167, 24 June 1973).

9

Environmental Impact Assessment

9.1

The environmental impact assessment procedure has only recently been introduced in Luxembourg law.

Under the terms of Article 1 of the Law of 17 July 1978 which amended the Law of 29 July 1965 regarding the conservation of nature and natural resources, all proposed developments or works outside built up areas,[1] 'which are likely to damage the environment as a result of their size or effect on the natural environment' can be made subject to an impact study. The decision to carry out an impact study is taken by the Minister responsible for the Water and Forestry Administration, at the suggestion of the Council of Government.

In addition, as already mentioned,[2] Article 6 of the Law of 16 April 1979 regarding dangerous, dirty or noxious establishments states that a summary assessment of the possible effect on the environment can be required for all industrial, craft or commercial establishments, whether public or private, and all manufacturing installations or processes, the existence, operation or bringing into service of which could result in danger or inconvenience, especially to the environment.

Finally, it should be noted that the judicious application and interpretation of the legislation on land-use planning (the Law of 12 June 1937 regarding the planning of towns and other major built up areas, and the Law of 20 March 1974 concerning land-use planning) also enables the environmental impact assessment procedure to be used and introduced into the planning process.

9.2

There are no precise indications in Luxembourg legislation regarding the elements involved in impact assessment.

The Minister of the Environment has issued the following statement on the subject:

> Impact assessments constitute a technical instrument of knowledge at the disposal of the decision process.

> What are the characteristics of the environmental impact assessment procedure and what are its consequences?

> Firstly, it obliges the public authorities and industry to assess the consequences on the quality of life and the natural environment of all measures taken or proposed at national or international level. Secondly, it should lead to the provision of reports on environmental impact which can be integrated into the administrative approval procedures for physical planning documents, regional and other development plans, industrial and infrastructural investment plans and even draft laws.

> The function of these reports would not be simply to ascertain that the proposed development conforms to the requirements in force with regard to the environment, but also to evaluate the global impact on ecological systems and on the quality of life of the population involved. For example, the direct consequence of the installation of a new industrial establishment can be air and water pollution problems or solid waste pollution problems, and it may in addition occupy a plot of land which would otherwise have been put to a different use. Indirectly, an impact can be felt through increased charges for the transport infrastructure of the region as a result of deliveries of raw materials and finished products, and also as a result of the creation of additional housing for the new workforce, etc. The purpose of an environmental impact report is thus to provide an overview of all the effects of the creation of this installation, whether long-term or short-term, direct or indirect, whenever important effects are involved.

> By its very comprehensive nature the assessment process requires impact studies or reports to be prepared in close collaboration with all parties involved who are able to supply assessment reports: firstly, the various administrative authorities responsible for the different environment sectors (air protection, water protection, land protection, etc.) which specify the constraints to be observed by the proposed works within their area of responsibility; secondly, the

351

individuals concerned who express their opinion on the impact of the works on their quality of life. This impact is often difficult to measure because of its subjective nature (what value should be given to the effects of noise, the increase of urban congestion, the dividing of a community, etc.) and in order to do so, a knowledge of the sensitivity of the individuals involved is essential. In addition to providing information, the fact of involving all the different authorities with sectoral responsibility for environmental matters in the assessment procedure can promote the coordination of authorisation procedures on matters which are often widely dispersed among too many different authorities. A coordinating action of this kind would greatly contribute to a reduction in the delays inherent in the authorisation of works of this kind.[3]

9.3

In practice, projects for installing new industries which will have an effect on the environment are effectively subject to an assessment of their impact on the environment.

The Minister of Energy had an impact study carried out on the installation of a nuclear power station.

An impact study was carried out by the Minister of the Environment on the subject of the motorway infrastructure in the Alzette Valley.

Notes

1. In the meaning of the Law of 27 July 1978, a built-up area consists of a group of at least five houses used in a permanent manner for human habitation, and situated within a radius of 100 m.
2. See Chapter 2.
3. Minister of the Environment, *Exposé budgétaire*, November 1979, p. 5–7.

10

Compensation for Damage to the Environment

10.1

In Luxembourg Law, as well as in Belgian Law, civil responsibility on the subject of pollution is based on the theory of fault.[1] The polluter is responsible for damages caused according to the principles contained in Article 1382 onwards of the Civil Code. The following must therefore be proved:

(a) the fault;

(b) the damage;

(c) causation.

10.2

The following constitute fault, unless there is a justifiable reason:

(a) the violation of a legal or regulatory provision or infringement of technical requirements;

(b) actions contrary to the obligation for due care and attention;

(c) the abuse of right.

10.3

In a ruling given on 10 November 1971, the Superior Court of Justice upheld the responsibility of successive polluters:

If a watercourse is no longer able to purify itself because, from time to time, it receives new harmful discharges, all those who have contaminated the watercourse are responsible for the damage caused.[2]

10.4

With regard to justifiable reasons—necessity, *force majeure,* acts by a third party—the rulings of the Superior Court of Justice of 23 December 1963 and 18 January 1964 decided that a period of abnormal drought does not constitute a case of *force majeure* for the polluter and that torrential rain does not constitute a state of necessity.[3]

10.5

With regard to the theory of 'the abuse of right', it would appear that Luxembourg jurisprudence takes a somewhat restrictive view. On 26 June 1979 the Superior Court of Justice noted that

> The owner of land with the right over its surface and sub-soil can use the water in a layer existing in the ground, whether it is stagnant or flowing, and regardless of the inconveniences experienced by the owners of lower lying lands.

> To this end, he can carry out all excavation works he considers necessary to find a water source, even if these excavation works result in cutting off the supply streams of a source which rises on lower lying land.

> Furthermore, he may make use of a well on his land in order to extract water, even if by this groundwater offtake he diminishes the supply streams of a source which rises on lower land.

> *The only exception to this ruling is when the owner of the higher ground has acted* without personal necessity or usefulness *but with the sole intention of causing harm.*[4]

10.6

As in Belgian Law, an action can be brought before the civil tribunals in the absence of any fault or defect, within the framework of the theory of private nuisances.

Article 544 of the Civil Code protects equally the rights of all owners. It follows that the right to property which permits its use and disposal in the absolute manner will be limited, not only by the restrictions provided in the law, but also by the need to respect the property of others.

The rights of the owner of a new building are limited by the rights of the property of the neighbours. While, in consequence, the normal inconveniences resulting from neighbourliness are to be tolerated without giving rise to compensation, inconveniences which exceed the limits of reciprocal tolerance between neighbours oblige the owner to repair the damage he has caused by his action. In each case, the court will make the overall assessment as to whether or not the normal inconveniences between neighbours have been exceeded.

In particular, compensation is due in matters relating to construction the moment a direct relationship between cause and effect is established between the new building and the damage suffered by the neighbour, provided that the damage is serious and exceeds the level of usual damage caused by building works carried out in the vicinity.[5]

10.7 THE RIGHT TO CLAIM COMPENSATION

In implementation of common law, the right to civil action is recognised for all physical and legal persons (corporations) harmed by an infringement.

Damage to the environment constitutes personal damage as soon as harm has been done to the physical wellbeing or property of the victim.

Similarly, an association can claim injury to its moral or material property as personal damage: e.g. an association of fishermen.[6]

Finally, the new Law of 27 July 1978 concerning the protection of the natural environment has introduced 'the associational action'. In Article 25 it states:

> Those associations of national importance with statutes published in the *Mémorial* appendices, which have carried out their statutory activities for at least 3 years in the area of the protection of nature and the environment can be the subject of Ministerial approval. Associations approved in this way can be called upon to participate

in the actions of public organisations which aim to protect nature and the environment.

In addition, these associations can exercise the rights recognised to civil parties with regard to acts which constitute an infringement in the meaning of this Law and which result in direct or indirect harm to the collective interests which they aim to protect.

10.8

Article 21 of the Law of 16 May 1929 concerning the care, maintenance and improvement of watercourses enables the communes to act through civil action to obtain compensation for any act liable to result in damage to watercourses.

In the absence of action by the communes, the Government can appoint a special commissioner to act in their name.

10.9 THE RESPONSIBILITY OF THE PUBLIC AUTHORITIES

The general rules concerning civil responsibility—Article 1382 onwards of the Civil Code—are applicable to the public authorities.[7,8]

10.10 COMPENSATION FOR DAMAGE

Right of damages is granted for injury suffered, if compensation in kind proves impossible.

The only anti-pollution law which contains a specific provision on the subject is the Law of 16 May 1929. In the event of an infringement of this Law, the judgement must require that the head of the business convicted compensates for the infringement within a certain period to be specified in accordance with Article 19, paragraph 2.

This compensation must extend exclusively to the damaging consequences produced by the actions of the offender, but it is not up to the tribunals to specify the method to be used in the future by the offender to purify the waste-waters produced in the operation of his industry.[9]

10.11

It is the duty of the president of the area tribunal—the civil judge in pleas of urgency—to bring about the cessation of acts which, in an intolerable and manifestly illegal manner, prevent a party from exercising or enjoying an incontestable right. Orders given on summary trial *(les ordonnances de référé)* are implemented provisionally, without prejudice to the principal.

Notes

1. A particularly well qualified author makes the following comment on the subject:

   It is a solution which is explained in the context of the individualistic liberalism of the 19th century. The essential subject of the rules of civil responsibility is the protection of the freedom of the individual and of the property of each individual.

   (R.O. Dalcq, 'La responsabilité civile et pénale du pollueur en droit belge, in *Les aspects juridiques de l'environnement*, Namur, 1975, p. 45.)

2. Unpublished decision, quoted by R. Everling, 'La responsabilité civile et pénale du polleuer en droit luxembourgeois, in *Les aspects juridiques de l'environnement*, Namur, 1975, p. 125.
3. See Chapter 5, note 8.
4. Superior Court of Justice, 26 June 1979, *Pas. lux.*, XXIV, 312.
5. Superior Court of Justice, 29 January 1963, *Pas. lux.*, XIX, 71.
6. Superior Court of Justice, 1 July 1964, unpublished, quoted by R. Everling, *op.cit.*, p. 125.
7. See Chapter 12, Belgium.
8. In a ruling issued on 31 October 1973, the area tribunal of Luxembourg decided that the Water and Forestry Administration was responsible, under the terms of Article 1 of 7 April 1909, under the authority of the Government, for the administration and supervision of the woods of the communes; it was therefore the responsibility of its agents to carry out detailed investigations of the depth of the roots of trees in the proximity of road works. The failure to carry out this obligation constitutes a fault and enables the State to be found responsible (*Pas. lux.*, XXII, 519).
9. Superior Court of Justice, 26 May 1951, *Pas. lux.*, XV, 154.

11
Conclusions

In the Grand-Duchy of Luxembourg environmental law has developed considerably over the last few years. While there are a few problems of coordination among the various public authorities, as a result of the small size of the country and the homogeneity of the population, it is true to say that a unification of environmental policy is actively sought.

On the organisational level, a national committee for the protection of the environment was created recently at governmental level to ensure the coordination of environmental protection measures. On the level of the administrative services, the Administration of the Environment was created.

In this way, an appropriate technical and administrative framework has been created enabling the most suitable preventive and remedial measures to be taken on environmental matters.[1]

With regard to industrial nuisance, the new Law of 16 April 1979 relating to *dangerous, dirty and noxious establishments* has taken into account the lessons of the past in protecting henceforth not only 'neighbourhood' interests but also the new interests involved in the protection of nature and the environment.

The various Grand-Ducal regulations issued in implementation of the outline Law of 21 June 1976 relating to the fight against *atmospheric pollution* appear to be producing very positive results. The controls carried out by the monitoring networks—the national network for sulphur fumes, the dust-measuring network at Esch/Alzette and Differdange, and the sampling installation at Luxembourg town for measuring lead levels in the air—show that the limits proposed by the European Communities are being observed.

The application of regulations relating to *noise control* issued in implementation of the Law of 21 June 1976 results in regular intervention by the Noise Service of the Administration of the Environment, consisting partly of acoustic measurements to verify conformity with the regulations and partly of an analysis of the situation with a view to the installation of a new establishment.

According to the 1980 budgetaty report of the Minister of the Environment, there are at present three problems:

(a) the incorporation of an automatic noise limiting device in the amplification system is essential to ensure the maximum sound level of 90 dB (A) required under Article 2 of the regulation concerning noise levels for music;

(b) an important part of the possible nuisance caused by discotheques arises from the use of cars and motorbikes by patrons of these establishments; it is therefore necessary, for all future establishments, for an impact study to be carried out on the establishment itself and of the parking facilities provided for the vehicles of patrons;

(c) in comparison with Belgium, the Grand-Duchy does not have adequate legal means to intervene against nuisances arising from racing vehicles and training vehicles used on private property.

As far as *waste-waters* are concerned, the figure of 85% biological treatment was exceeded during 1980. The treatment of watercourses with successive hydraulic installations with a view to improved self-purification is considered to be one of the most important contibutions to land development policy in the Grand-Duchy.

On the basis of a global analysis of the existing legislation, the text of a general draft bill concerning the management and the protection of watercourses is now being prepared.

The general revision of the legislation and its adaptation to the new requirements should result in a law dealing with all the aspects of water policy, and covering surface waters as well as groundwaters. It will contain a great number of provisions concerning, amongst others:

the objective of the law and the legal definitions;

authorisation procedures;

a coordinating body;

works for conservation of navigable and non-navigable watercourses;

protection of groundwater;

the quality of bathing water;

treatment of waste-waters;

implementation of the 'polluter pays' principle;

control organisations;

penal provisions;

abrogatory provisions.

In this way, the text will allow the coordination of sectoral approaches to the different problems to achieve a single horizontal and integrated approach. The new law will constitute an important step forward in the realisation of environmental legislation; it will create a basic framework for the authorisation procedure and for specific regulations to be completed in the future.[2]

Legislation concerning *waste management* was completely revised in 1980; in addition, a Grand-Ducal regulation relating to the disposal of toxic and dangerous waste is in preparation.

Finally, the exemplary attitude of the Grand-Duchy of Luxembourg with regard to the introduction into internal law of the directives issued by the Council of the European Communities should be noted.

In spite of the serious economic difficulties resulting from the energy crisis in which it finds itself—as do all industrial societies today—the Grand-Duchy of Luxembourg has begun to deploy its resources in a constructive and integrated policy of environmental protection, giving priority to an improved quality of life, in other words, towards a balanced economy, without waste of resources. It is a subject of astonishment as well as, and above all, of admiration.

Notes

1. Budgetary Report by the Minister of the Environment, November 1980, p. 3.
2. Bedgetary Report by the Minister of the Environment, November 1981, p. 3.

Classified Index*

* References are to section numbers.